RECENT RESEARCHES IN CORROSION EVALUATION AND PROTECTION

Edited by **Reza Shoja Razavi**

Recent Researches in Corrosion Evaluation and Protection
Edited by Reza Shoja Razavi

Published by InTech
Janeza Trdine 9, 51000 Rijeka, Croatia

Publishing Process Manager Martina Durovic

Technical Editor Teodora Smiljanic

Cover Designer InTech Design Team

Additional hard copies can be obtained from orders@intechopen.com

Recent Researches in Corrosion Evaluation and Protection, Edited by Reza Shoja Razavi
p. cm.
ISBN 978-953-307-920-2

For EU product safety concerns:
IN TECH d.o.o., Prolaz Marije Krucifikse Kozulić 3, 51000 Rijeka, Croatia,
info@intechopen.com or visit our website at intechopen.com.

Meet the editor

Dr Reza Shoja Razavi is an Assistant Professor of Materials Engineering at Malek Ashtar University of Technology. He graduated from Malek Ashtar University of Technology (1st Rank in BSc program of Materials Engineering, 1999), Tarbiat Modaress University (1st Rank among all graduates of the MSc program of Materials Engineering, 2001), and Isfahan University of Technology (1st Rank among all graduates of the PhD program of the University, 2007). He specializes in Surface Engineering with primary interests in corrosion science. He is a permanent member of the Iranian Corrosion Association, Iranian Society of Surface Science and Technology and National Elite Foundation. He has directed ten research projects and more than ten MSc theses, published more than 80 technical papers and two books, and delivered eight courses to BSc, MSc and PhD students. He has been the recipient of many honors and awards for his excellence in both teaching and research.

Contents

Preface

Corrosion is expensive and can be hazardous. It is costly to replace and repair equipment, structures, and other industrial items that have been damaged as a result of corrosion. In order to understand the mechanisms of corrosion, solve the corrosion problems, and improve the corrosion resistance of materials, the use of modern evaluation and protection techniques is essential. Therefore no significant progress can be achieved in this area without the use of modern evaluation methods combined with electrochemical techniques. It is the purpose of this book to present and discuss the recent methods in corrosion evaluation and protection.

This book contains six chapters. The aim of Chapter 1 is to describe the cataphoretic electro-deposition process and to demonstrate, using some practical examples, that Electrochemical Impedance Spectroscopy can be a very useful tool to provide a complete evaluation of the corrosion protection properties of electro-coatings. Chapter 2 presents results of studies of materials degradation: localized corrosion of metals and the interaction of carbon monoxide with nickel-iron alloys from experimental electrochemical tests and theoretical calculations. Evaluation of the corrosion process of stainless steels in natural seawater and atmospheric corrosion of copper in a marine tropical-humid climate are presented and discussed, making emphasis on the electrochemical techniques used. Chapter 3 deals with the presentation of the corrosion and corrosion prevention of the aluminium alloys by organic coatings and inhibitors. Industrial applications and common corrosion form of different Al alloys are carried out in this chapter. Chapter 4 addresses the new method of pigment preparation named "Core-shell" that can lead to the production of new pigments with improved properties different from each of its individual components, overcoming their deficiencies; and consequently changing their efficiency of protection when applied in paints. The effectiveness of plasma-deposited films on the improvement of carbon steel corrosion resistance is discussed in Chapter 5. The corrosion resistance is determined by electrochemical impedance spectroscopy. Chapter 6 deals with the conjugation of carbon nanotubes with organic-inorganic hybrid to prepare hybrid coatings that combine high anti-corrosion efficiency with elevated mechanical resistance. The corrosion protection efficiency was investigated by electrochemical impedance spectroscopy (EIS) and potentiodynamic polarization curves.

I hope the book will be very useful for corrosion engineers, students and all those interested in corrosion evaluation and protection. I am very grateful to InTech for giving me the honor of being the editor of this book and Malek Ashtar University of Technology, who gave me moral support. Finally, I am especially indebted to my wife, Samira Tabatabaee, for her mental support.

Reza Shoja Razavi
Department of Materials Engineering
Malek Ashtar University of Technology
Shahinshahr, Isfahan
Iran

1

Use of Electrochemical Impedance Spectroscopy (EIS) for the Evaluation of Electrocoatings Performances

Marie-Georges Olivier[1] and Mireille Poelman[2]
[1]*University of Mons, Materials Science Department, Mons,*
[2]*Materia Nova Research Centre, Mons,*
Belgium

1. Introduction

Thanks to their barrier properties against corrosive species, organic coatings are often used to protect metals against corrosion. In the automotive industry, cathodic electrocoating is widely used as a primary layer coating in the corrosion protection system [1-3]. This deposition method has many advantages including high throw power, high corrosion protection and coating transfer coefficient (>95%), auto-limitation of the coating thickness, environmentally friendly due to an aqueous suspension medium and an easy industrial automation [4, 5]. This coating can also be applied on each metal composing the car body.

Corrosion protection is guaranteed only if good adhesion properties are attained between the metallic substrates and the coating. Enhanced adhesion can be achieved by the use of an appropriate surface preparation prior to coating (etching, polishing, etc.) or with a good pre-treatment which can also provide additional corrosion protection and adherence [6-9]. So an appropriate combination of surface preparation, pretreatment and coating provides increased durability of the protective system.

As high durability systems are continuously developed, short-term test methods are required to evaluate the corrosion resistance of paint/metal systems and decide among coatings designed for long-term durability. The principle of usual ageing tests is based on the application of specific stresses (temperature, humidity, salts, UV light) at higher levels than in natural exposure to induce accelerated deterioration of the system. A first objective of an accelerated test is to cause the degradation of the coating or its failure in a shorter time period than under natural conditions without changing the failure mechanisms. So far a direct correlation between natural degradation and the weathering device currently being used is not clear, therefore accelerated tests are generally only used for comparative purposes. The degradation of the coating can be obtained using different ageing tests such as immersion in electrolyte, continuous salt fog or SO_2 exposure. Cyclic corrosion tests combining different kinds of exposures (humidity, salt fog, drying steps) are also increasingly used [10-14]. These methods are based on the principle that corrosion can only occur if electrolyte and oxidant species are present at the metal surface. The increase of temperature allows accelerating the transport of oxygen and electrolyte through paint and initiating the corrosion reactions. For some accelerated tests, use is made of scratched panels to simulate coating damage and delamination. This type of acceleration is essentially related to the damage of protective properties of the coating/pre-treatment/metal system and not to the barrier properties of the coating. As the protective properties of electrocoating systems are continuously improved ageing tests necessitate increasing times before any visual observation of the initiation of degradation is possible.

Electrochemical Impedance Spectroscopy is a powerful tool which was widely used in the last decades to characterize corrosion processes as well as protective performances of pre-treatments and organic coatings [15-28]. This electrochemical technique is not destructive and can consequently be used to follow the evolution of a coated system exposed to an accelerated ageing test and provide, in short time, information about the corrosion kinetics.

In the automotive industries, the most challenging failure modes needing to be detected and evaluated are: the barrier properties of the electrocoating during immersion, thermal cycles or salt spray test; the behavior of a scratched sample in terms of extension of the delaminated area for different kinds of exposures and the specific corrosion at the edges.

The aim of this chapter is to describe the cataphoretic electrodeposition process and to demonstrate by some practical examples that Electrochemical Impedance Spectroscopy can be a very useful tool to provide a complete evaluation of the corrosion protection properties of electrocoatings.

2. Cataphoretic electrocoating

The electrodeposition process was used for the first time in 1963 by Ford in USA to paint spare parts. The process was based on the anodic electrodeposition. In 1967, PEUGEOT was the first car manufacturer to employ this process to coat the whole body car by anaphoresis. In 1978, Chrysler-France in Poissy was the first European line to coat body cars by using cationic paint. Since then, all the manufacturers protect vehicles against corrosion by cataphoresis [1].

In cationic coatings, positively charged coating particles dispersed in an aqueous solution are electrophoretically attracted to a substrate, which is the cathode of the electrolytical cell. The increase of pH due to water reduction at the cathode induces the electrocoagulation of the coating on the substrate. In the case of anaphoresis, the substrate is the anode and the coating binder is charged negatively. In both processes, thermosetting binders are used. In this chapter, only the cataphoresis process will be described.

Cationic electrodeposition presents numerous advantages: self-limitation of the coating thickness, low water permeability (dense network), good adherence and adhesion, high throw power, automation ability, low loss in products, low pollution level and high corrosion protection.

2.1 Principle

In cataphoresis, the metallic substrate is linked to the negative terminal of an electrochemical cell and immersed in the coating bath during more or less two minutes. Under the effect of the electrical field induced by application of a high voltage difference (in the range of 200 V to 500 V) between the piece (cathode) and the counter-electrode in a low conductivity paint bath, the positively charged coating particles (grafting of amine groups), move towards the cathode by electrophoresis.

The electrochemical reactions are the following:

- **At the cathode (piece to coat):** water electrolysis with hydrogen production and OH^- formation. This reaction provokes a local increase of pH which neutralizes the NH_3^+ groups fixed on the binder. The particle becomes insoluble in water and settles on the metallic substrate. This step corresponds to the electrocoagulation of the paint.

$$4H_2O + 4e^- \rightarrow 2H_2 + 4OH^-$$

- **At the anode (counter-electrode):** water is electrolyzed with oxygen evolution and H^+ production. This increase of H+ ions raises the bath conductivity.

$$2H_2O \rightarrow O_2 + 4H^+ + 4e^-$$

At last, the water contained in the coating layer is expulsed by electro-osmosis (water movement under the action of electrical field through the porous layer).

In brief, the coating cathodic electrodeposition is the result of four elementary phenomena (Figure 1):

1. Electrophoresis;
2. Water electrolysis;
3. Coating electrocoagulation;
4. Water electro-osmosis.

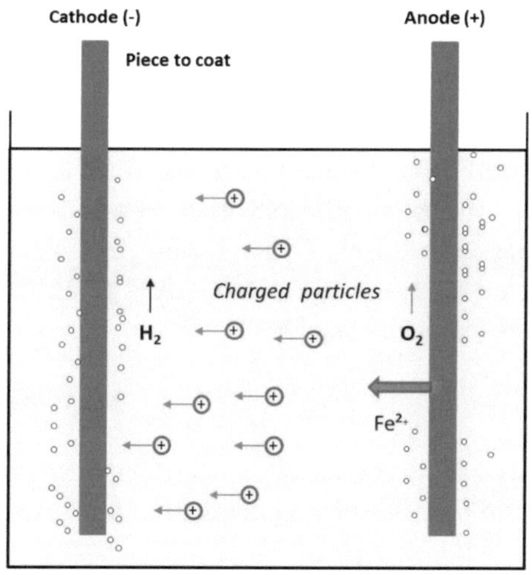

Fig. 1. Schematic representation of cataphoretic deposition.

The resins used in the formulation of cationic automotive primers (E-coat) are based on epoxy resins. The epoxy resin is reacted with an aminoalcohol, such as diethanolamine (DEA) to obtain a resin with amine and hydroxyl groups. The resin product reacts with an isocyanate half-blocked with an alcohol as 2-ethylhexylalcohol or 2-butoxyethanol for example in order to make a self cross-linkable resin [29]. Catalyst and pigment dispersions (titane dioxide or carbon black) are added. In aqueous medium the pigments have a charge which is insufficient to form a homogeneous suspension. The ionized binder surrounds them as a protective colloid and insures suspension homogeneity and stability. The pigment diameter must be lower than one micron.

The amine groups are neutralized with a volatile carboxylic acid such as acetic or formic acids before dispersion in water. The neutralization rate plays an important role in the stability and deposit process of the coating. In order to determine this parameter, a titration of a known amount of paint by HCl and KOH allows calculating the numbers of alkaline (Basic Meq) and acid (Acid Meq) milliequivalents, respectively.

The neutralization rate is given by equation (3):

$$T_n = \frac{Alkaline\,Meq}{Acid\,Meq + Alkaline\,Meq}$$

It is usually in the order of 40-50%.

The binder is insoluble in water but can be dispersed thanks to the ionization of fixed groups. This ionization is carried out by using an acid according to the following reaction step:

$$L\text{-}(NH_2)_n + m\ HA \longrightarrow L\text{-}(NH_2)_{n\text{-}m} - (NH_3^+)_m + m\ A^- + m\ H_2O$$

The electrocoagulation step is the opposite of the solubilization operation. The neutralized form of the binder being insoluble in water and hydrophobic, the disposal of liquid paint can be done by water rinsing. This property is employed to recover and recycle the paint excess. The water is also expulsed during the electro-osmosis step. This phenomenon allows diminishing the energy consumption during the curing. The water weight percent in the layer is around 10-15% before baking.

After electrocoagulation, the film has to be reticulated by curing at controlled temperature. This operation confers the definitive properties to the film: adherence, hardness, corrosion resistance. During this step, the blocked isocyanate reacts with a hydroxyl group of the epoxy resin to form a urethane cross-link. Older E-coat primers contained basic lead silicate as a catalyst, which interacted with the phosphate layer to enhance adhesion. Lead-free E-coats have replaced the older formulations [30, 31].

2.2 Industrial process

The installations (Figure 2) are preceded by a surface treatment area (phosphating or alternative pretreatments) and followed by a curing area (around 30 min at 130-180°C).

The industrial process is composed of the following steps:

- Coating bath ;
- Systems to maintain bath stability ;
- Rinsing installation ;
- Power supply ;
- Conveyer.

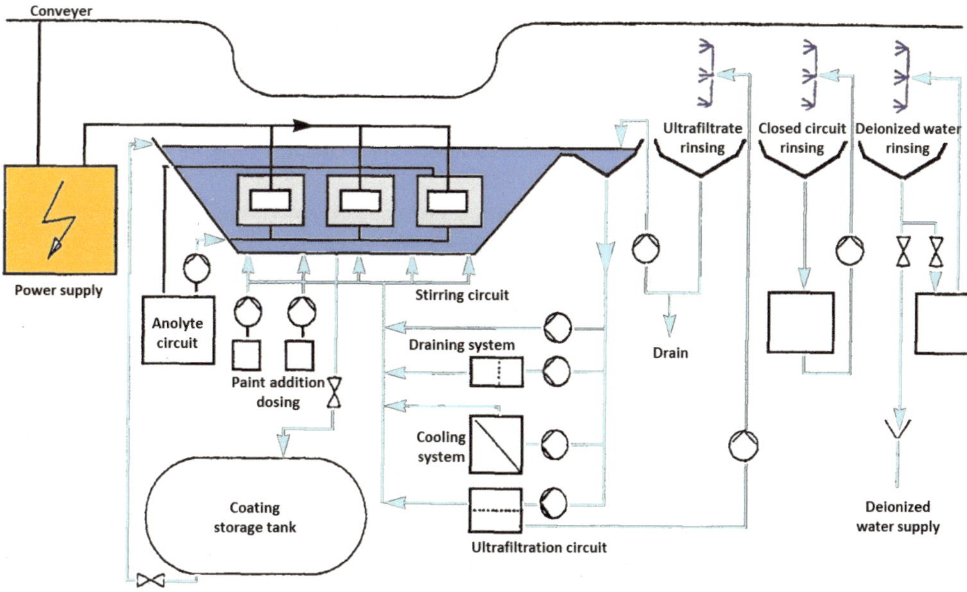

Fig. 2. Schematic representation of industrial process of cataphoresis.

Bath coating

The capacity and the bath dimensions are depending on the size of the pieces to coat. The tank is in PVC or in steel covered by corrosion protection coating. An overflow, built on the side of the tank, supplies the different circuits.

Fluids circulation

- **Coating circulation:** the stirring of the coating is carried out by an external coating circuit supplied by a pump. In order to have a sufficient stirring, the pumped coating must be reintroduced by injection nozzles located in the bottom or on the lateral faces of the tank. This circulation prevents the coating flocculation.
- **Filtration circuit of the paint:** Elimination of the impurities carried into the bath is achieved by continuous filtration.
- **Cooling system:** the quality of the coating can only be insured and reached in a narrow range of temperature. The heating of the bath is due to the current flow, bath stirring or due to the heat brought by the pieces coming from the surface treatment.
- **Ultrafiltration circuit:** Despite the fact that the film thickness is controlled by the applied voltage, there is always during emersion a supplementary entrainment of paint which has to be eliminated by rinsing. The paint contained in rinsing water can be recovered and recycled by ultrafiltration. This technique consists in a separation on a membrane. The matters flow through the membrane is achieved by pressure application depending on the average size of the membrane pores. The ultrafiltration membranes have an average porosity comprised between 0.001 and 0.1 µm and work under pressure from 1 to 10 bar. The retained matters, also called concentrates, are binder macromolecules, pigments and fillers. The matters which cross the membrane are the ultrafiltrate composed of dissolved compounds. The dry content is around 1%. The ultrafiltrate is used in the rinsing system. The concentrate is recycled towards the coating bath.

- **Draining circuit of impurities:** The bath is contaminated by impurities coming from surface treatments and altering the paint. These impurities, non-stopped by membrane, are continuously eliminated thanks to a draining system located in the ultrafiltration circuit. This draining is compensated by addition of deionized water. Depending on the quality and the nature of rinsing after surface treatment, order of 0.001 to 0.005 liter of ultrafiltrate per painted square meter is rejected.
- **Anolyte circuit:** The acid needed for binder neutralization (produced at the anode) is not eliminated after electrocoating and enriches the bath. Being harmful the acid excess is extracted by electrodialysis on cationic membrane. This method consists of protons transport through a selective membrane under the effect of an electrical voltage and their concentration in an anodic compartment of the electrodialysis cell where they are extracted and replaced by deionized water. This circuit allows maintaining the bath pH at 6-7.
- **Circuit of paint addition:** the paint deposited on the pieces must be compensated by addition of a preparation concentrated in binder, pigments and fillers. These compounds are either premixed outside the tank with the bath paint or directly introduced from the stirring circuit.

Rinsing system

During emersion of the pieces, non-coagulated paint is carried by capillarity and retention in the pores. This paint amount can reach from 25% to 45% of the consumed paint and must be removed by rinsing before curing due to its different characteristics. The pieces rinsing is done in three steps. During the first step, the film is rinsed with ultrafiltrate. For the second rinsing, a closed circuit is used. For the high quality coatings, a third rinsing is performed with deionized water. Before curing, the pieces are dried by air flushing.

Power supply

The electrocoating installations require a power supply in continuous current with voltage in the range of 200 – 400 V (1 to 10 mA/cm^2).

Conveyer

Conveyer allows the pieces transport but also constitutes the negative terminal of the generator. The conveyer design must take into account the immersion duration, positioning and number of pieces to coat.

2.3 Advantages, drawbacks and challenges

Effect of application parameters

The main parameters needing control during electrodeposition are voltage, bath conductivity and temperature [32].

The rate of deposition is strongly affected by applied voltage: the higher voltage induces a faster deposition. The conveyer is designed to obtain a coating in 2 to 3 min at a voltage comprised between 225 and 400 V. The high voltage increases the driving force for electrophoretic attraction of the particles to the cathode and allows an adequate covering of the confined areas. The first areas covered are the edges of the metallic substrate due to their highest current density. The electrical resistance rises with the film thickness reducing the rate of electrodeposition. There is a limiting film thickness beyond which the coating deposition stops or at least becomes very slow. When the edges are coated, outer flat surfaces of the body car are coated, followed by recessed and confined areas. For corrosion protection, it is needed to have the entire surface of the metal coated and to coat the furthest confined areas within the 2 to 3 minutes dwell time in the tank. The deposition in the recessed area explains the high throw power of electrocoating which increases with the applied voltage and the duration in the bath. However, if the applied voltage is too high, a film rupture on the outer surfaces will be observed due to current flow leading to local generation of hydrogen under the film. The generation of gas bubbles blowing out through the film induces film defects. Throw power is also affected by the bath conductivity: a higher conductivity induces a greater throw power. Nevertheless, there is a limitation, an increase of bath conductivity modifies the conductivity of the film (presence of soluble salts) induces an increase of the ionic strength and so a loss of the bath stability due to decrease of the zeta potential of the double layer capacitance at the interface between the charged particles and the bath. A compromise must consequently be reached. The bath conductivity is from 1200 to 1800 μS/cm.

The properties of the film are also strongly depending on the temperature which must be controlled in a very narrow range, typically 32 to 35°C.

Advantages and drawbacks

Electrodeposition is a highly automated system and requires only one operator. The transfer coefficient is higher than 95%. However, the financial investment of the automated line is high, limiting applicability of these lines to large production processes. The eletrodeposition unit is the most expensive equipment in an auto assembly plant.

Solvent content of E-coats is relatively low, so VOC emissions are limited and fire risk reduced. The complete coverage of surfaces is another advantage. Even if differences are observed in film thicknesses, the inner areas having thinner layer than the exposed face areas, the entire surface will be protected. Objects with many edges can be better coated by electrodeposition than by any other painting technique.

Uniform thickness can be a problem, especially with relatively highly pigmented primers: as the applied coating follows closely the surface contours of the metal, a rough metal will give a rough primer surface.

The paint films are relatively thin, varying from 15 to 30 μm, depending on coating composition and application parameters. The substrate must be conductive and only the first layer can be applied by electrodeposition.

These coatings present very high performances if well applied and require efficient methodology to evaluate their corrosion properties after application and ageing. The corrosion protection of this layer necessitates the development of electrochemical techniques allowing a rapid detection of microdefects, loss of barrier properties, delamination propagation on scratched samples,... This information should ideally be available in a short time, before the defects are observed by visual or optical inspections. In the following paragraph, the electrochemical impedance spectroscopy is described for the evaluation of the main degradation risks of electrocoating during lifetime.

3. Electrochemical impedance spectroscopy

This part of the chapter will describe the principle, the interpretation of impedance spectra by using the raw parameters and the electrical equivalent circuits. The evaluation of performances of electrocoatings will be discussed and illustrated by some practical examples.

3.1 Principle

Electrochemical impedance spectroscopy is a non-stationary technique based on the differentiation of the reactive phenomena by their relaxation time. The electrochemical system is submitted to a sinusoidal voltage perturbation of low amplitude and variable frequency. At each frequency the various processes evolve with different rates, enabling to distinguish them.

A weak amplitude sinusoidal perturbation is generally superimposed to the corrosion potential or open circuit potential:

$$\Delta U = |\Delta U| \sin\omega t \quad \text{with } \omega = 2\pi f$$

where f is the frequency (Hz) of the applied signal.

This perturbation induces a sinusoidal current ΔI superimposed to the stationary current I and having a phase shift φ with respect to the potential:

$$\Delta I = |\Delta I| \sin(\omega t - \varphi)$$

These values can be represented in the complex plane:

$$\Delta U = \Delta U_{re} + i\Delta U_{im}$$

$$\Delta I = \Delta I_{re} + i\Delta I_{im}$$

The complex impedance is defined as:

$$Z = \frac{\Delta U}{\Delta I} = Z_{re} + iZ_{im}$$

The impedance can also be represented by a modulus $|Z|$ and a phase angle shift φ:

$$|Z| = \sqrt{Z_{re}^2 + Z_{im}^2}$$

$$tg\varphi = \frac{Z_{im}}{Z_{re}}$$

The impedance data can be represented in two ways:

- Nyquist spectrum: $-Z_{im}$ as a function of Z_{re}
- Bode spectrum: log $|Z|$ and phase angle φ as a function of log f

The electrochemical measurements are generally carried out using a conventional three-electrode cell filled with the electrolytic solution (Figure 3): a working electrode (the sample under study), a counter electrode (often a platinum grid or plate) and a reference electrode (such as Ag/AgCl/KCl sat.).

The exposed surface area must be accurately determined and should be high enough when coating capacitance needs to be evaluated.

The impedance measurements are performed over large frequencies ranges, typically from 100 kHz to 10 mHz using amplitude signal voltage in the range of 5 mV to 50 mV rms. The amplitude is strongly depending on the studied system. For an electrocoating system, a classical value of 20 mV rms is chosen to characterize intact coatings. However, for EIS measurements on scratched samples, the response of the exposed metal being dominant, a perturbation of maximum 5 mV rms is used. The EIS spectra can be acquired using the combination of a potentiostat with a frequency response analyser or with a lock-in amplifier.

As the temperature may strongly influence the kinetics of water or oxygen diffusion, the corrosion rates and the mechanical properties of the film, the measurements are preferably carried out at controlled temperature.

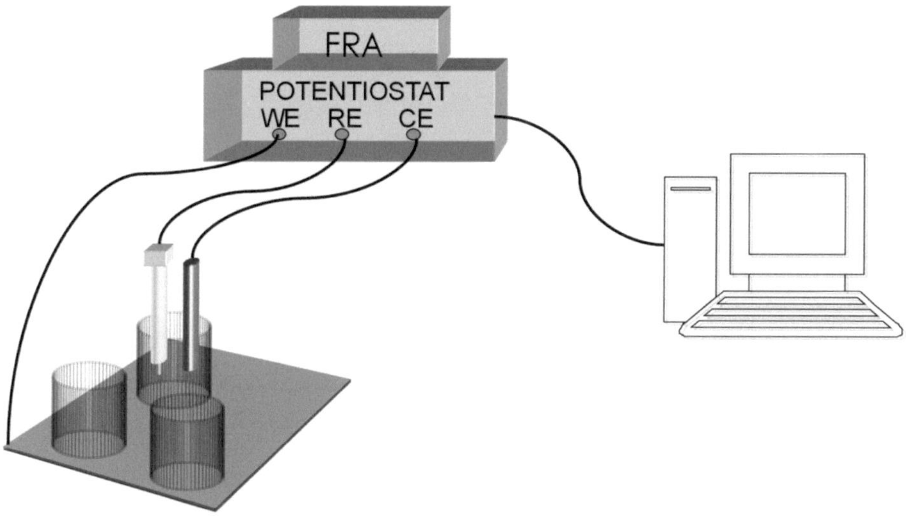

Fig. 3. Schematic representation of the electrochemical cell

3.2 Electrical equivalent circuit

The interpretation of impedance data is generally based on the use of electrical equivalent circuits representative of the electrochemical processes occurring at the sample/electrolyte interface. These circuits are built from the appropriate combination of simple electrical elements (capacitors, resistors,…).

An intact coating behaves as a dielectric and can be represented by a capacitor. When in contact with an electrolyte, the coating starts to absorb water and the electrolyte enters the pores of the coating. The electrical equivalent circuit describing this system is represented in Figure 4. While entering the pores, the electrolyte causes a decrease of the pore resistance R_p which can be considered as initially infinite.

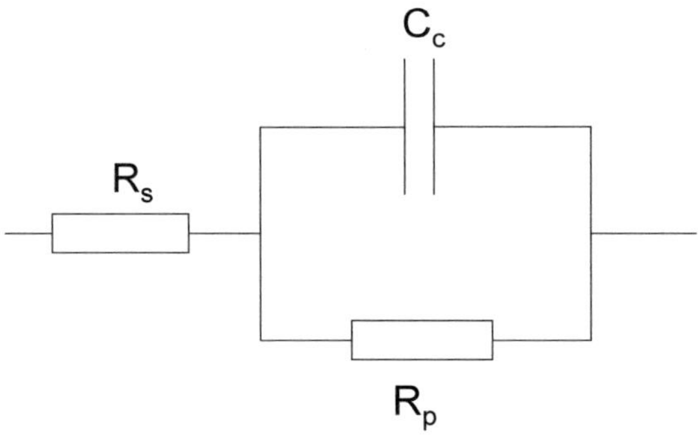

Fig. 4. Electrical equivalent circuit of an intact coating in contact with an electrolyte; R_s is the electrolyte resistance, C_c the coating capacitance and R_p the pore resistance.

Once the corrosion reactions start at the metal/electrolyte interface under the coating or at the base of the pores electrical elements related to the newly created interface have to be included in the equivalent circuit. This is illustrated in Figure 5 in which a circuit describing the exposed metal/electrolyte interface is added to the coating electrical elements. This circuit consists of the double layer capacitance and an electrical element describing the electrochemical reactions at the metal/electrolyte interface.

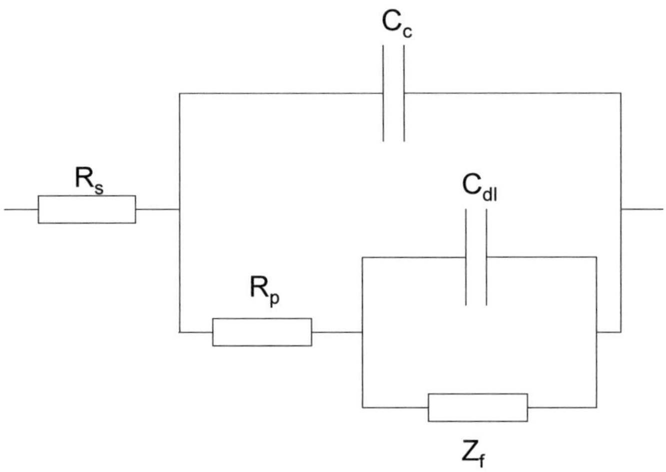

Fig. 5. Electrical equivalent circuit of a coating in contact with an electrolyte (degradation in process). R_s is the electrolyte resistance, C_c the coating capacitance, R_p the pore resistance, C_{dl} the double layer capacitance, Z_f an electrical element representing the electrochemical reactions of the metallic substrate in contact with the electrolyte [33].

The fitting of the impedance data to the circuits of Figures 4 or 5 allows obtaining the electrical parameters describing the coating:

- C_c, the coating capacitance defined by

$$C_c = \frac{\varepsilon_0 \varepsilon_r A}{d}$$

where ε_0 is the vacuum permittivity or the permittivity of the free space, ε_r is the relative permittivity or coating dielectric constant, A the coating surface area and d its thickness.

The dielectric constant of a typical polymeric material is about 3 – 8 and that of water being at 78.5 at 25°C. During water absorption by the coating, its dielectric constant increases with the resulting increase of the coating capacitance (A and d considered as constant). It is possible to estimate the amount of absorbed water. The most common model to calculate the volume fraction of absorbed water was developed by Brasher and Kingsbury [34] and is given by:

$$\phi = \frac{\log(C_t/C_0)}{\log(\varepsilon_w)}$$

where C_t is the coating capacitance at time t, C_0 the capacitance of the 'dry' coating and ε_w is the water dielectric constant.

The use of this equation is however restricted to some limitations:

- The increase of the coating capacitance should only be due to water absorption
- Water is uniformly distributed inside the coating
- The water uptake remains low and no swelling occurs
- No interaction between water and the polymer may occur

- R_p, the pore resistance defined as

$$R_p = \frac{\rho d}{A_p}$$

where ρ is the electrolyte resistivity in the pores, d the pore length (~coating thickness) and A_p the total pore surface area.

R_p decreases as the electrolyte penetrates the coating and fills the pores. The decrease of Rp with time may be related to the increase of A_p which can be explained by an increase of the number of filled pores or an increase of their area if delamination occurs.

- C_{dl} : the double-layer capacitance which is proportional to the active metallic area A_w (area in contact with the electrolyte).
- Z_f which in the simple case of a charge transfer–controlled process can be replaced by R_{ct}, the charge-transfer resistance inversely proportional to the active metallic area.

In practical, the measured impedance spectra may differ from ideal or theoretical behaviour. The loops (or time constants) do not show a perfect semi-circle shape in Nyquist representation. This non-ideal behaviour may arise from coating heterogeneities as roughness, inhomogeneous composition,… In such a case the coating cannot be described by a simple capacitor. This one is generally replaced by a constant phase element (CPE) whose impedance is given by:

$$Z_{CPE} = \frac{1}{Y_0}(i\omega)^{-n}$$

n accounts for non-ideal behaviours: when it equals to 1, the CPE is a pure capacitance and when it equals zero, the CPE is a pure resistance.

Misinterpretation of coating evolution may arise from an erroneous impedance data fitting. Consequently, it is sometimes better to restrict the data interpretation to simple parameters as the global resistance of the system represented by the low frequency impedance modulus ($|Z|_{0.01Hz} \sim R_s+R_p+R_{ct}$) and the coating capacitance values obtained from high frequency impedance modulus [35].

The coating capacitance can indeed be determined from the impedance modulus at a fixed frequency (10 kHz for example) and can thus be calculated from:

$$C_c = \frac{1}{2\pi 10^4 |Z|_{10kHz}}$$

3.3 Barrier properties

The barrier properties of an electrocoating are maintained during ageing if a capacitive behavior is maintained over the whole frequency range. Due to the high performances of cataphoretic coatings, the immersion time must be very long before observing changes in the EIS spectra. So some procedures to accelerate the ageing can be used in order to be able to differentiate the behavior of two different systems. Two methods will be discussed in this chapter, the use of salt spray and AC/DC/AC tests.

3.3.1 Immersion test

One example to illustrate the evaluation of the barrier properties is to investigate the EIS response of an experimental epoxy coating cataphoretically deposited on a 6016 aluminium alloy (typically used in the automotive industry). Two pretreatments are compared: acid etching and a commercial Zr/Ti conversion coating. The evolution of the coating properties is evaluated by EIS after different immersion times in NaCl 0.5 M electrolyte. Systems having different pretreatments are compared in terms of barrier properties, water uptake, and apparition of a second time constant in the EIS spectra. Figures 6a and b show the evolution as a function of time in the NaCl solution of the impedance modulus versus frequency of coated samples (Bode-modulus plots). Different stages can be distinguished: capacitive (C), mixed capacitive and resistive (CR) and resistive (R) behaviours. In the early times of immersion, the coating acts as a barrier against water and electrolyte (stage C). The coating behaves as a dielectric and the resulting impedance modulus logarithm varies linearly as a function of the frequency logarithm. The loss of barrier properties corresponds to the penetration of water and electrolyte through the pores and defects of the coating up to the metal. At this stage (CR), the low-frequency modulus progressively decreases reflecting the decrease of the pore resistance (R_p). The time at which the low-frequency modulus starts to decrease varies slightly with the surface preparation prior to coating: an average of 40 days for non pre-treated samples (NP) (Figure 6a) and 50 days for Zr/Ti pre-treated samples (ZT) (Figure 6b). For both surface preparations, a rapid decrease of the low-frequency modulus is observed from the moment the coating lost its barrier properties. Anyway, this experience evidences the importance of the surface pretreatment on the barrier properties of the cataphoretic electrocoating. This can be explained by a more homogeneous surface on the pretreated substrate showing an uniform electrochemical activity during cathodic electrodeposition. Non pretreated samples are generally rather heterogeneous. Hydrogen evolution reaction may thus vary from one point to another with the consequent risk of appearance of coating micro-defects after coating curing.

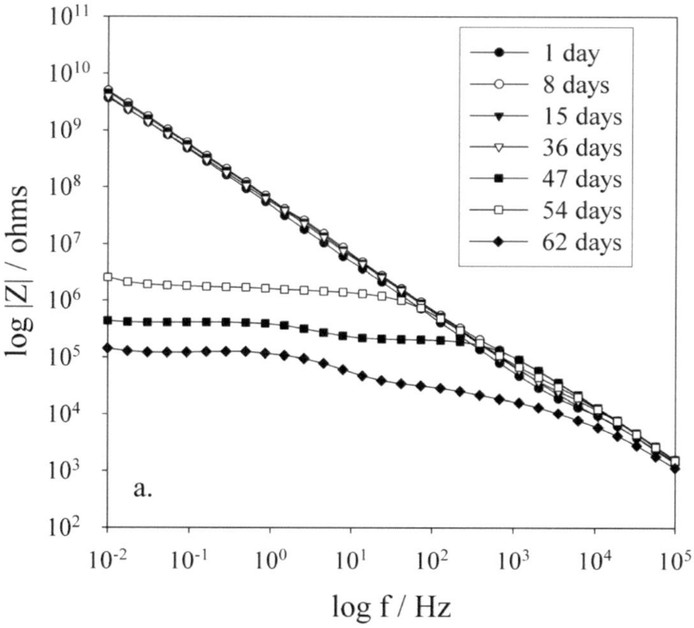

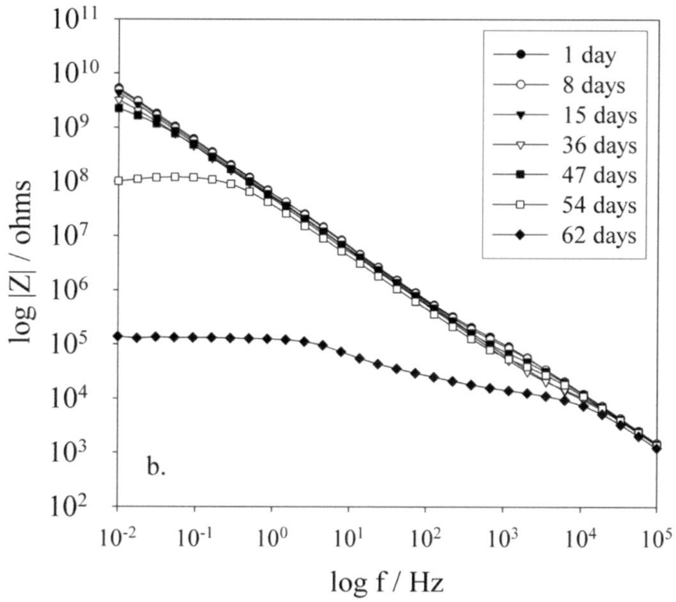

Fig. 6. Bode-modulus plots for different immersion times in NaCl 0.5 M of electrocoated aluminium samples without pre-treatment (a) and with a Zr/Ti pre-treatment (b) [36].

Once the electrolyte reaches the base of the pores, corrosion of the metal may start. Corrosion may significantly affect the adhesion of the coating by the increase of pH accompanying the oxygen reduction or by the presence of corrosion products. The loss of adhesion will cause an increase of the active metal surface area so that a second loop or time constant (R_{ct} C_{dl}) appears in the Bode-diagrams. The second time constant starts to be visible after 47 days and 62 days for NP and ZT samples respectively.

Coated samples were also exposed to salt fog. In that case, the samples were periodically removed from the salt spray chamber and transferred into the electrochemical cell filled with NaCl 0.5M. After an EIS measurement the samples were immediately returned to the salt fog chamber so that the time spent out of the chamber was as short as possible (maximum one hour).

As with the continuous immersion test, the three stages of degradation are also observed with the salt spray test combined with EIS. There is a rather good correlation between the results obtained with both tests. Indeed, the EIS spectra obtained as a function of immersion time or salt fog exposure show that the Zr/Ti pre-treatment enhances the corrosion resistance of the coating. Moreover for ZT samples, after 40 days of exposure to salt fog the total impedance is still high as for the immersion test after the same testing period. The time at which non pre-treated samples lose their barrier properties is somewhat shorter for salt fog exposure than for the immersion test. This difference can be accounted for by the higher exposure temperature and the higher oxygen concentration due to continuous aeration in the case of salt fog exposure. At the end of both tests, no visible signs of deterioration were detected while significant changes in the EIS response occurred.

The pore resistance was determined by fitting the impedance diagrams with the electrical equivalent circuit model of Figure 5. In agreement with the observed decrease of the low-frequency impedance modulus, R_p shows an important decrease as a function of exposure (Figure 7), especially for samples NP which give R_p values below 10^7 ohms cm^2 after 40 days of exposure. The pore resistance is higher for ZT samples than for NP samples suggesting that the pre-treatment could also have a beneficial role on the barrier properties of the coating.

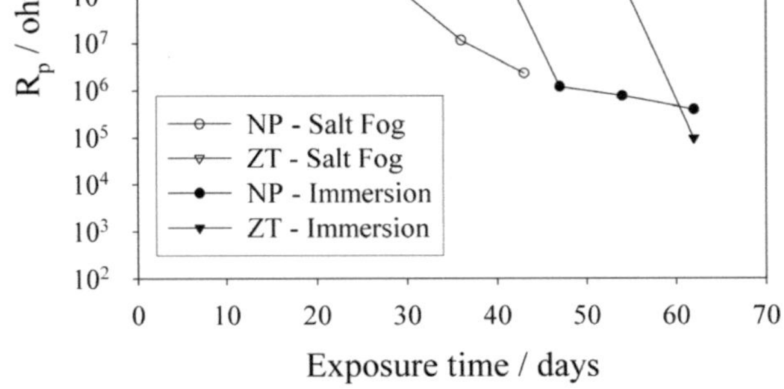

Fig. 7. Pore resistance as a function of exposure time to salt fog or to an immersion test for NP and ZT samples [36].

The coating capacitance values reflect water and electrolyte ingress in the coating. The C_c values are almost constant versus exposure time for samples NP and ZT exposed to both tests, meaning that during the period from 1 to 60 days there is probably no significant water absorption by the coating. Cataphoretically deposited coatings are known to be rather impermeable to water so that there is only a slight absorption of water by the coating at the early moments of contact as illustrated in Figure 8. The volume fraction of water absorbed by the coating can be estimated from Brasher and Kingsbury and is about 0.8% which is very low.

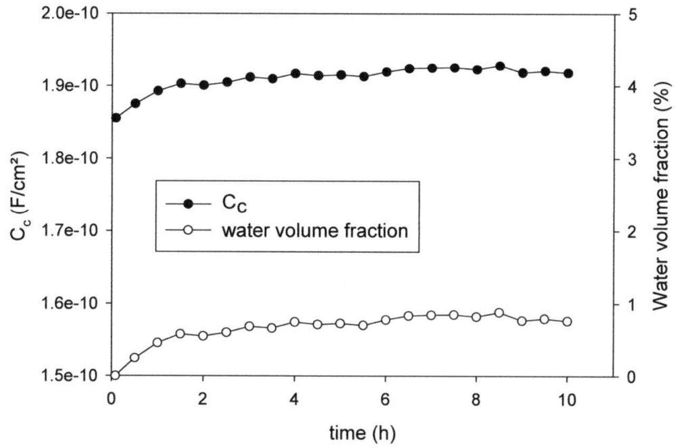

Fig. 8. Coating capacitance as a function of exposure time in NaCl 0.5M for a ZT sample.

3.3.2 AC/DC/AC cycles

The ac/dc/ac procedure was developed in order to assess the anti-corrosive properties of a coating in a very short time [37-41]. This procedure consists of a combination of cathodic polarization (dc) and EIS measurements (ac). After a first ac measurement at the open-circuit potential, the sample is treated for a short time by a constant cathodic potential (dc). Typically, during the dc period, the tested sample is cathodically polarized at -3V/ref during 2 h followed by a 3 h relaxation time until it recovers a new steady state. These steps are repeated by means of programmed cycles until the loss of the coating protective properties is observed in the ac spectrum. The evolution of the impedance spectrum is generally attributed to both coating degradation due to the decrease of pore resistance and to delamination process, which is accelerated by OH^- production at the metal surface during cathodic polarization.

As the density of pores or micro-defects reaching the metallic substrate decreases with the coating thickness, using thinner coatings may give rise to even more pronounced degradation and provide useful information in shorter time. In the following example the ac/dc/ac procedure was performed with two different electrocoatings (A and B) differing by their content in plasticizer. These coatings were applied with different thicknesses (14, 17 and 20 μm) by changing the electrodeposition time and maintaining the application voltage constant. An example of the resulting impedance spectra is presented in Figure 9 for a Zr/Ti pretreated sample with coating A applied with 14 μm thickness.

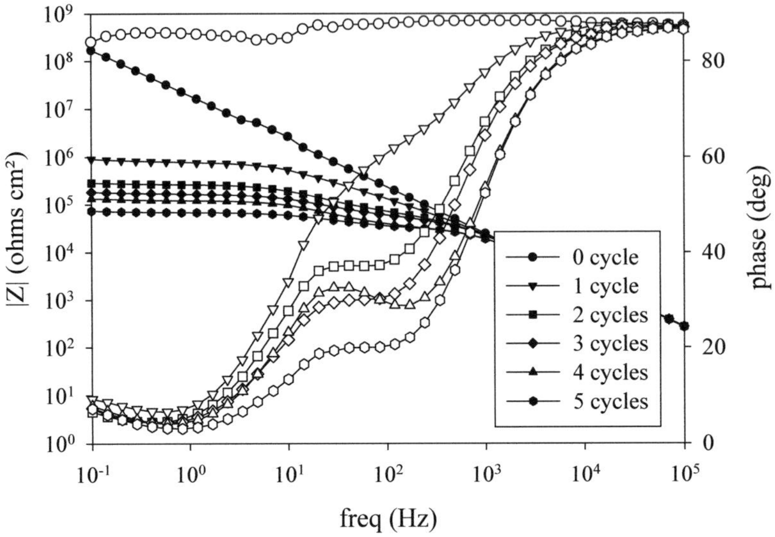

Fig. 9. Bode (modulus and phase) plots for an electrocoated sample (coating A) undergoing ac/dc (-3V/ref)/ac cycles. Electrolyte : NaCl 0.5M, exposed surface area : 4.9 cm^2.

Though the initial barrier properties are high and the behavior purely capacitive, the total impedance of the system strongly decreases after one cycle. Afterwards the barrier properties continue to decrease gradually with the number of cycles. With low thickness, the decrease of R_p is even more pronounced. The faster decrease of the barrier properties when the coating thickness is reduced is probably due to a higher density of pores reaching the metallic substrate.

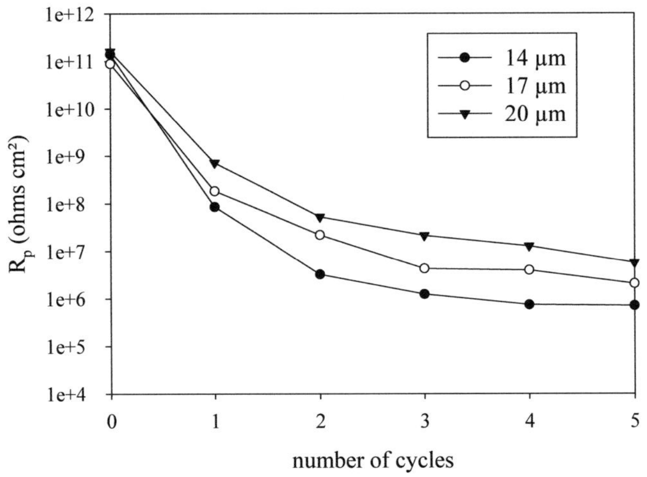

Fig. 10. Pore resistance as a function of the number of ac/dc cycles for coating A with a Zr/Ti pretreatment.

The presence of a distinguishable second time constant from the first ac/dc cycle allows the determination of the electrical parameters related to the metal/electrolyte interface. These parameters were obtained by fitting the impedance spectra with the equivalent circuit of Figure 5 for the different coating thicknesses. An increased electrochemical activity at the metal/electrolyte interface results from the application of ac/dc/ac cycles. As a consequence the double layer capacitance (C_{dl}) shows a rather important increase with the number of cycles (Figure 11).

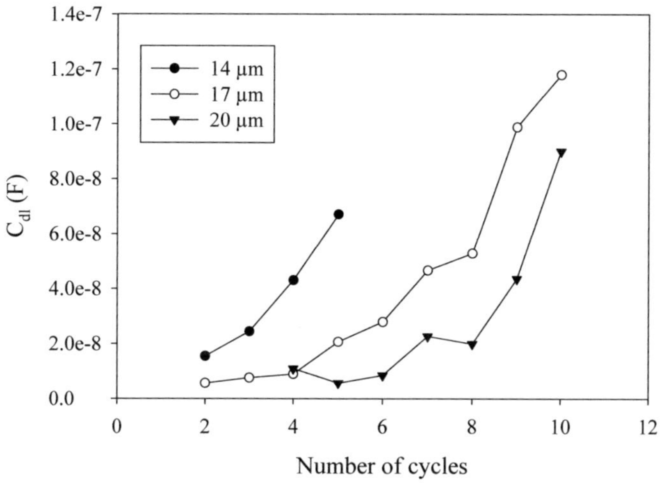

Fig. 11. Double layer capacitance as a function of the number of cycles as determined from fitting the impedance spectra with the electrical equivalent circuit from Figure 5. Electrocoating A applied with different coating thicknesses on etched aluminium samples (without pretreatment).

Applying a cathodic polarization induces a fast penetration of the electrolyte through the total thickness of the coating. Delamination occurs in the pore regions and favors the blisters formation. The increase of the double layer capacitance is less pronounced and appears later when the coating thickness is higher as illustrated in Figure 11. For a same number of cycles, the double layer capacitance is indeed higher for a lower coating thickness. This can be accounted for by the higher density of pores reaching the metal leading to a total higher value of the active metallic surface area. Smaller thicknesses consequently result in a rapid start of the corrosion process and in a consequent adhesion loss of the coating from the substrate.

Applying an appropriate pretreatment as Zr/Ti leads to higher barrier properties and to a better resistance to cathodic delamination as accounted for by the smaller double layer capacitance values determined with ZT samples whatever the electrocoating applied (Figure 12). A comparison between two coatings (A and B) differing by their content in plasticizing agent is also possible on the basis of the AC/DC cycles as illustrated in Figure 12. The coating with the higher content in plasticizer shows lower barrier properties and a high sensitivity to cathodic delamination accounted for by the important increase of C_{dl} with the number of cycles especially on NP samples. The addition of plasticizing agent leads to better rheological properties during curing but is also responsible of a less curing density and thus to higher porosity and water permeability. The barrier properties of such coatings may however be enhanced through an appropriate adaptation of the coating parameters (such as voltage for example). AC/DC cycles thus offer a rapid method to discriminate "bad" coatings and to optimize the application parameters.

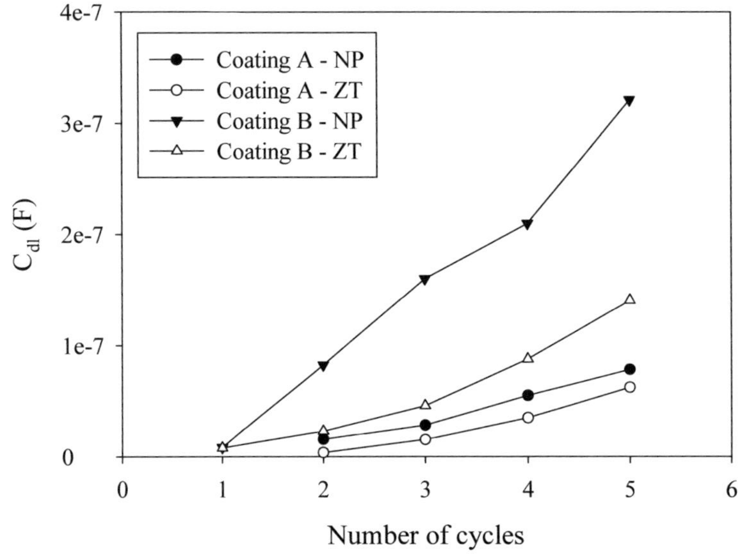

Fig. 12. Double layer capacitance as a function of the number of cycles as determined from fitting the impedance spectra with the electrical equivalent circuit from Figure 5. Electrocoating A and B applied on etched aluminium samples (NP) and on Zr/Ti pretreated samples (ZT). Coating thickness: 14 μm.

3.4 Evaluation of delaminated area from a defect

The most common way to evaluate the delamination of a coating is to scratch the coating, reaching the metallic substrate, to expose the samples to an accelerated ageing test and to evaluate after a determined exposure time the total delaminated area from the scratch. It is however also possible to follow the evolution of the delamination process with EIS measurements on scratched samples [21, 22, 42-45]. In that case, the evaluation of the delaminated area is based on the determination of the wet area where the corrosion reactions take place. Nevertheless, in order to obtain quantitative values of this surface by EIS spectra the corrosion products coming from the corrosion phenomenon must be dissolved before EIS determination. The apparition of these corrosion products may also be avoided by cathodic polarization. In each case the operating parameters must be well controlled to avoid a subsequent degradation or induce a degradation process not representative of a natural ageing. These parameters were investigated to identify a new electrochemical tool to evaluate the improvement of the interface stability of metal/electrocoating in order to identify the sensitivity to filiform corrosion of electrocoated aluminium alloys, the electrocoating coverage of steel edges and the delaminated area during a salt spray test.

Aluminium alloys are known to be particularly susceptible to filiform corrosion which is a specific delamination which occurs under atmospheric conditions with high relative humidity (50-90%). This specific type of delamination is an anodic undermining driven by a differential aeration cell created between the head front of filament (anode) and the defect (cathode). This corrosion phenomenon can be revealed by EIS on scratched samples [21, 22, 44]. In the following example, the electrocoated aluminium alloy samples were scratched with a cutter reaching the metallic substrate. The linear defect produced was 2±0.02 cm long and 40 µm width with an area of about 0.8 mm square. The width and area of the defect were controlled with the help of an optical microscope. The procedure used to initiate (1 hour in the HCl vapours) and propagate filiform corrosion was the same as that adopted in the ISO/DIS 4623 standard with a shorter exposure time in the humid chamber. The climatic conditions during exposure were 82 ± 3% relative humidity and 40°C ± 2°C. The samples were then analysed by EIS at room temperature in 0.1 M Na_2SO_4 acidified at pH ranging from 1 to 3 by adding sulphuric acid. Different immersion times in the electrolyte solution were explored. The analysis of the EIS data is based on the fitting of the spectra with the electrical equivalent model of Figure 5. However in the presence of a macroscopic defect as that obtained with a scratch, the time constant associated with the coating is generally shifted towards higher frequencies than those conventionally explored. The low/mid frequency time constant accounts for the electrochemical processes occurring at the exposed metallic substrate. C_{dl} and Rct are two parameters used to specify the delamination or filiform corrosion of the coating. The choice of an acidified electrolyte solution (0.1 M Na_2SO4 at pH =2) allows to dissolve the corrosion products formed under the coating during exposure to the humidity chamber. The immersion time in the electrolyte is also an important parameter. Actually, the immersion time has to be long enough to dissolve the corrosion products formed during the test. However, too high immersion times may be accompanied by the growth of new corrosion products due to the reaction of the metal with the testing electrolyte. Figure 13 illustrates the effect of immersion time in the testing electrolyte (sodium sulphate at pH 2) for an electrocoating applied on etched aluminium (without pretreatment) and exposed 48h to the standard filiform corrosion test (82% RH and a temperature of 40°C). After immersion for 1 h in the testing electrolyte, two time constants can be distinguished in Bode-phase diagram. The time constant observed at high frequency can be assigned to the presence of corrosion products as discussed in ref 46-48. This means that the corrosion products formed under the coating during exposure to humidity are not completely dissolved before EIS measurement. The low frequency time constant attributed to the corrosion process is displaced to lower frequencies indicating a slowdown of the process which can be explained by the contribution of diffusion in the electrochemical process occurring at the metal/electrolyte interface [49]. Longer immersion times in the electrolyte allow the dissolution of the corrosion products present in the defect as only one time constant is detected from 8h of contact with the electrolyte.

In order to enhance the dissolution of the corrosion products in a shorter time, measurements were also carried out at pH = 1. The impedance data in the Bode phase representation obtained after 48 h in the climatic chamber, 4 h in the test solution at pH = 1 are shown in Figure 14. To get a better understanding of the influence of pH, measurements were also made at pH = 3. By acidifying the electrolyte solution, the corrosion products are dissolved within shorter immersion times and thus, the time constant due to corrosion products in the impedance data disappears. On the contrary, after 4 h of immersion in the electrolyte solution at pH = 3, the time constant associated with the electrochemical reaction can be explained by a higher contribution of diffusion process than at pH = 2. At pH=1, the observed time constant is only related to the electrochemical reaction at the metal/electrolyte interface in the scratch and inside the filaments.

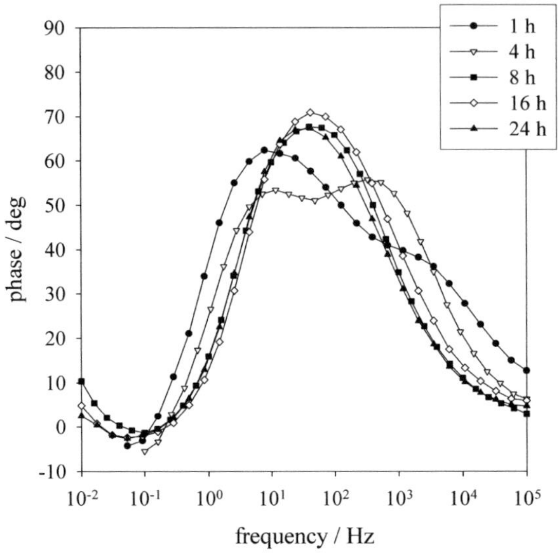

Fig. 13. Bode-phase diagrams for different immersion times in the sulphate electrolyte solution at pH 2. Data obtained with aluminium samples etched and electrocoated without any pre-treatment; exposure time in the humidity chamber: 48 h [48].

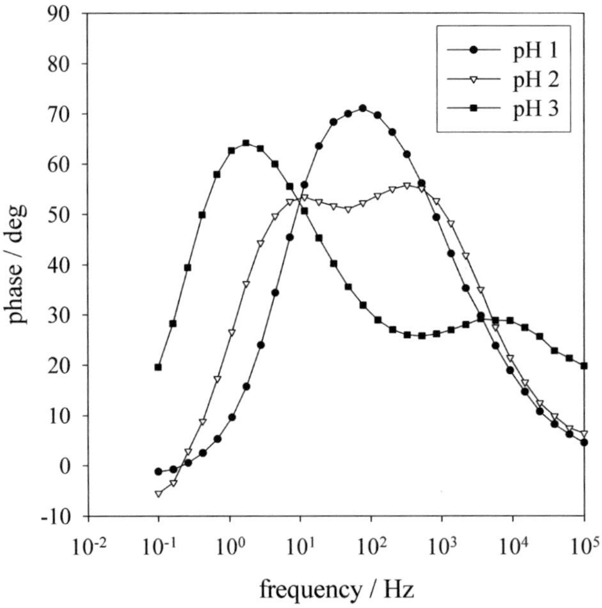

Fig. 14. Bode-phase diagrams at different pH of the sulphate electrolyte solution. Aluminium samples coated without pre-treatment; data obtained after 48 h of exposure to humidity chamber and 4 h of immersion in the electrolyte solution [48].

This simple procedure which was initially developed to the study of filiform corrosion on coated aluminium may also be used to evaluate the delaminated area of scratched samples exposed to any type of accelerated ageing test. In the following example, EIS was performed on electrocoated aluminium 6016 samples exposed for maximum 10 days to a neutral salt fog. After exposure, the scratched samples were assembled in a three-electrode electrochemical cell filled with Na_2SO_4 0.1 M at pH 1. The impedance measurements were carried out after 1 h, time necessary to reach stationary conditions (stable corrosion potential). The resulting impedance spectra only show one time constant corresponding to the second time constant of electrochemical circuit shown in Figure 5.

C_{dl} values were determined by fitting the impedance spectra with one time constant electrical model describing the electrochemical reactions occurring at the exposed metal surface. The C_{dl} values were then divided by the double layer capacitance of the bare metal to give the evolution of the active metal area as a function of exposure (Figure 15). Higher

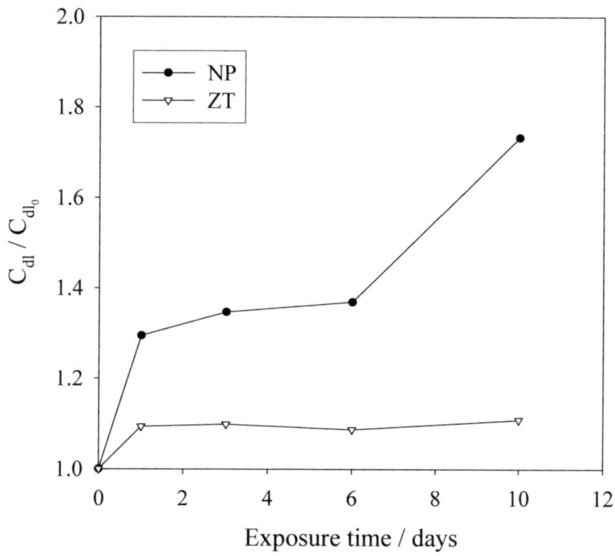

Fig. 15. Values of the double layer capacitance C_{dl} divided by the bare metal double layer capacitance C_{dl0} as a function of exposure to salt fog of scratched NP and ZT samples [36].

double layer capacitance values are observed for NP samples explained by the poor adhesion of the coating when the samples are not pre-treated before being painted. Moreover non pre-treated samples show a significant increase of the double layer capacitance as a function of the exposure time to salt fog and thus an increase of the delaminated area. For the Zr/Ti pre-treated samples, the values are constant during the first 10 days of exposure to salt spray. Accordingly, the tendency observed for C_{dl} values of both samples fully agrees with the visual observation of a development of filiform corrosion on non-pre-treated samples while Zr/Ti pre-treated samples were not degraded after 10 days of salt fog exposure. After this period, the solubilization of the corrosion products formed on non-pre-treated samples becomes harder. As a consequence, a rather accurate estimation of the extent of filiform corrosion or delamination can be only obtained for short exposure times. However in the present case 10 days of exposure is long enough to distinguish the behavior of the two types of samples (NP and ZT) and to observe the increase of the delaminated area on non pre-treated samples. Consequently, combining exposure to salt fog of intact and scratched samples and following their degradation by EIS allow the evaluation of the influence of the pre-treatment on the loss of barrier properties and on the coating adhesion to the substrate.

3.4.1 Edge corrosion

In spite of the ability of cathodic electrocoating to cover the totality of the car body the electrocoat paints are sensitive to edge corrosion [50]. The edge corrosion generally results from the absence of film or low film build at the edges responsible of a premature corrosion. The edge coverage is especially linked to the flow properties of the coating during the baking process. High-edge coverage coatings are developed by adding rheological agents to the coating formulation [51-53]. The edge coverage can also be improved by modifying the electrocoat application parameters as the voltage, thickness or by the addition of a resistance in the circuit [50]. At present, to characterize the protection offered by cataphoretic coatings against edge corrosion, use is made of knife blades of which edge has a known angle of 38°. After a close examination of the knife blades, these are coated and then undergone a 7 days neutral salt fog exposure (35°C, 5% NaCl) (NF X 41-002). After exposure the blades are rinsed with deionized water and dried. The corrosion is then characterized by numbering the rust spots that appeared on the edges. Though this quotation is generally made with the help of a microscope it is not easy to count every single rust spot [50, 54]. This method is thus highly time consuming, laborious and rather subjective since the results may depend on the operator. Different samples could be used instead of the knife blades as for example ultra-thin or perforated panels or the Volvo grooved steel cylinder. Such samples could improve the visual detection or differentiation of edge corrosion but would not accelerate the test.

Electrochemical methods could be used to characterize the edge-corrosion protection of electrocoated knife blades and to get a short time evaluation of the parameters of the coating deposited on the blade edge.

In the following example, knife blades covered with a 20 μm coating thickness (thickness measured on the flat part of the blade) were exposed to salt fog for 7 days. The number of rust spots observed by microscope is given in Figure 16. These values present a rather important dispersion which can be attributed to the difficult numbering of rust spots, especially when the coating is highly degraded. Despite this dispersion, it is clear that the coating containing 3.5% of rheological agent is less corroded than the one containing 1%. Thus, a slight increase in rheological agent content probably leads to a better edge coverage and a subsequent higher resistance to edge corrosion. The influence of the coating thickness was studied for the coating containing 1% of rheological agent. Knife blades covered with 5, 10 and 20 μm coating thickness (measured on the flat part) were exposed to salt fog for 7 days. As shown in Figure 16 increasing the coating thickness improves the resistance to edge corrosion since for a 20 μm thickness no rust spot was detected. However it is not possible to differentiate the 5 and 10 μm thicknesses on the basis of the salt fog exposure since the average number of rust spots are very close for these thicknesses.

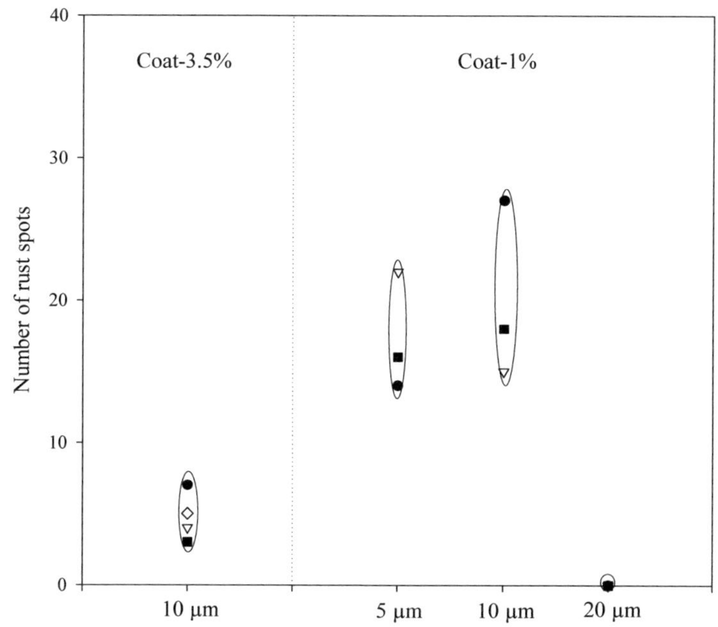

Fig. 16. Number of rust spots after 7 days of salt fog exposure

Different electrochemical methods can be used to characterize the blade edge/electrolyte interface. The idea is to find a test which could provide a rapid evaluation of the total exposed metallic area at the base of the pores or defects on the blade edge. Impedance measurements can provide this information. However, fitting the impedance spectra with the electrical equivalent circuit for an organic coating is not always possible. Another manner to obtain the exposed metallic area is to polarize the sample at a sufficient cathodic potential causing cathodic reaction as hydrogen evolution. The measured current will be proportional to the total exposed metallic area without being affected by the presence of corrosion products. To illustrate this, cathodic polarization measurements were performed on coated knife blades. The current measured at -3V/Ag/AgCl/KCl(sat) is used to distinguish different coated systems. In the present case the influence of the coating thickness of a coating containing 1% of rheological modifier is investigated. As illustrated in Figure 17, the current density at -3V/Ag/AgCl shows an important decrease with coating thickness. The lowest current density was measured for a coating thickness of 20 µm for which the exposed metal area is thus the smallest in agreement with the smallest number of rust spots observed after salt fog exposure. Moreover there is a noticeable difference in the current densities between 5 and 10 µm coating thicknesses, for which no distinction was possible on the basis of the salt fog exposure test. The results obtained by electrochemical measurements thus show a good correlation with the salt fog exposure test. However the time necessary to obtain the same information is very short since a cathodic polarization measurement takes about half an hour while the salt fog test necessitates 7 days of exposure.

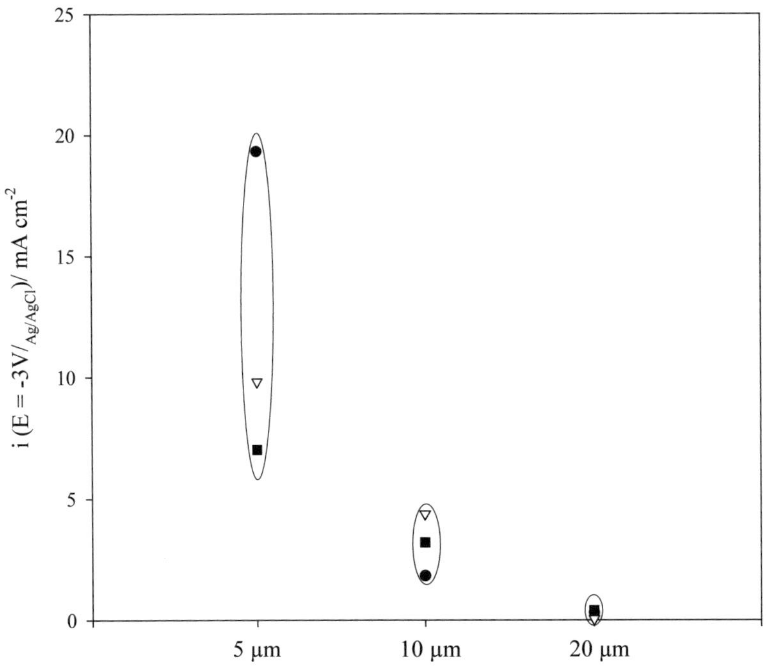

Fig. 17. Current density at a cathodic potential of -3V/Ag/AgCl for different coat-1% thicknesses (measured on the flat part) deposited on knife blades.

4. Conclusions

Cataphoretic electrodeposition is an industrial process which allows obtaining high coating performances responding to high quality requirements of automotive industry. Moreover as it is water-based and formulated without lead and limited VOC content this coating system has a low impact on the environment.

The important challenges in development of new coatings is to assure the adhesion and barrier properties with new environmentally friendly pre-treatments, to be applicable whatever the nature of the substrate, to cover all the car body parts even the recessed areas and the edges and to be scratch resistant. As showed by some examples, Electrochemical Impedance Spectroscopy is a powerful tool to evaluate in short times the behaviour of intact and scratched electrocoated samples. The EIS procedures can give more quantitative information compared to the classical ageing tests. However this technique needs an accurate control of the operating conditions. Furthermore the interpretation of the results needs an appropriate choice of the electrical equivalent circuits to describe the system under study.

5. Acknowledgments

This study was performed in the framework of the Opti2Mat project financially supported by the Walloon region in Belgium.

6. References

[1] Beck, Fundamentals of Electrodeposition of Paint, BASF, 1979, pp. 1–55
[2] T. Brock, M. Groteklaes, P. Mischke, *European Coatings Handbook,* Verlag, Hannover, 2000, pp. 279–285
[3] E. Almeida, I. Alves, C. Brites, L. Fedrizzi; *Prog. Org. Coat.* 46 (2003) 20
[4] K. Arlt, *Electrochim. Acta* 39 (1994) 1189
[5] L. Krylova, *Prog. Org. Coat.* 42 (2001) 119
[6] C.M Reddy, R.S Gaston, C.M Weikart, H.K Yasuda; *Prog. Org. Coat.* 33 (1998) 225
[7] C. A. Ferreira, S. Aeiyach, A. Coulaud, P. C. Lacaze; *J. Appl. Electrochem.* 29 (1999) 259
[8] M. Fedel, M.-E. Druart, M. Olivier, M. Poelman, F. Deflorian, S. Rossi; *Prog. Org. Coat.* 69 (2010) 118
[9] C.Wu, J. Zhang, *J. Coat. Technol. Res.* 7 (2010) 727
[10] G. P. Bierwagen, L He, J Li, L Ellingson, D.E Tallman; *Prog. Org. Coat.* 39 (2000) 67
[11] G.P. Bierwagen, "The Science of Durability of Organic Coatings—A Foreword." Prog. Org. Coat., 15 (1987) 179–185
[12] F.X. Perrin, C. Merlatti, E. Aragon, A. Margaillan; *Prog. Org. Coat.* 64 (2009) 466
[13] N. LeBozec, D. Thierry; *Materials and Corrosion* 61 (2010) 845
[14] A. W. Hassel, S. Bonk, S. Tsuri, M. Stratmann; *Materials and Corrosion* 59 (2008) 175
[15] L. Beaunier, I. Epelboin, J.C. Lestrade and H. Takenouti *Surf. Technol.,* 4 (1976) 237
[16] M.W. Kendig and H. Leidheiser.; *J. Electrochem. Soc.* 23 (1976) 982
[17] J.D. Scantlebury and K.N. Ho. *JOCCA,* 62 (1979) 89
[18] T. Szauer. *Prog. Org. Coat.* 10 (1982), 157
[19] F. Mansfeld, M.W. Kendig and S. Tsai. *Corrosion* 38 (1982) 478
[20] A. Amirudin, D. Thierry, *Prog. Org. Coat.* 26 (1995) 1
[21] F. Deflorian, L. Fedrizzi, *J. Adhesion Sci. Technol.* 13 (1999) 629
[22] D. Loveday, P. Peterson, B. Rodgers; *J. Coat. Tech.* 10 (2004) 88
[23] D. Loveday, P. Peterson, B. Rodgers; *J. Coat. Tech.* 2 (2005) 22
[24] F. Deflorian, S. Rossi; *Electrochim. Acta* 51 (2006) 1736
[25] K. Allahar, Q. Su, G.P. Bierwagen; *Prog. Org. Coat.* 67 (2010) 180
[26] S. Touzain ; *Electrochim. Acta* 55 (2010) 6190
[27] T. Breugelmans , E. Tourwé, J.-B. Jorcin, A. Alvarez-Pampliega, B. Geboes, H. Terryn, A. Hubin ; *Prog. Org. Coat.* 69 (2010) 215
[28] E.C. Bossert et al., U.S. patent 5,880,178 (1999)
[29] E.C. Bossert et al., U.S. patent 6,042,893 (2000)
[30] Z. W. Wicks, F.N. Jones, S. P. Pappas, D.A. Wicks, *Organic Coatings Science and Technology*, Third edition, John Wiley & Sons, (2007) 535
[31] F. Mansfeld, Electrochim. Acta 38 (1993) 1891
[32] D.M. Brasher, A.H. Kingsbury, *J. Appl. Chem.* 4 (1954) 62
[33] G.P. Bierwagen, D. Tallman, J. Li, L. He, C. Jeffcoate, *Prog. Org. Coat.* 46 (2003) 148
[34] M. Poelman, M.-G. Olivier, N. Gayarre, J.-P.Petitjean, *Prog. Org. Coat.* 54 (2005) 55
[35] M. Bethencourt, F.J. Botana, M.J. Cano, R.M. Osuna, M. Marcos *Prog.Org. Coat.* 49 (2004) 275-281
[36] S.J. Garcia, J. Suay *Prog. Org. Coat.* 59 (2007) 251-258
[37] M.T. Rodríguez, J.J. Gracenea, S.J. García, J.J. Saura, J. Suay *Prog. Org. Coat.* 50 (2004) 123-131
[38] S.J. García, J. Suay; *Progr. Org. Coat.* 66 (2009) 306-313

[39] K.N. Allahar, G. Bierwagen, V.J. Gelling ; *Corr. Sci.* 52 (2010) 1106-1114
[40] S. Gonzalez, M.A. Gil, J.O. Hernandez, V. Fox, R.M. Souto, *Prog. Org. Coat.* 41 (2001) 167
[41] J.M.I. McIntyre, H.Q. Pham, *Prog. Org. Coat.* 27 (1996) 201
[42] F. Deflorian, L. Fedrizzi, S. Rossi, P.L. Bonara, *Electrochim. Acta* 44 (1999) 4243
[43] M.L. Zheludkevicha, K.A. Yasakau, A.C. Bastos, O.V. Karavai, M.G.S. Ferreira; *Electrochem. Comm.* 9 (2007) 2622
[44] L. Fedrizzi, F. Deflorian, S. Rossi, *Benelux Metall.* 37 (1997) 243
[45] L. Fedrizzi, F. Deflorian, S. Rossi, P.L. Bonora, *Mater. Sci. Forum* 289–292 (1998) 485
[46] M.-G. Olivier, M. Poelman, M. Demuynck, J.-P. Petitjean, *Prog. Org. Coat.* 52 (2005) 263-270
[47] C.G. Oliveira, M.G.S. Ferreira, *Corr. Sci.* 45 (2003) 139
[48] N. Blandin, W. Brunat, R. Neuhaus, E. Sibille, *Proc. Eurocorr*, 2004
[49] Y.-B. Kim, H.-K. Kim, J.-W. Hong *Surf. Coat. Technol.*, 153 (2002) 284-289
[50] V.C. Corrigan, S.R. Zawacky, PPG Industries, Inc., "Cationic Microgels and Their Use in Electrodeposition", U. S. Patent 5,096,556, 1992
[51] D. Saatweber, B. Vogt-Birnbrich, *Prog. Org. Coat.* 28 (1996) 33-41
[52] M. Poelman, M.-G. Olivier, N. Cornil, N. Blandin, *Proc. Eurocorr*, 2006

2

Application of Electrochemical Techniques and Mathematical Simulation in Corrosion and Degradation of Materials

Jorge González-Sánchez, Gabriel Canto,
Luis Dzib-Pérez and Esteban García-Ochoa
*Centre for Corrosion Research,
Autonomous University of Campeche,
Mexico*

1. Introduction

The tropical climate prevailing at Yucatan Peninsula in Mexico is characterized by permanently high temperatures and relative humidity with considerable precipitation, at least during part of the year. A high corrosion rate of metals is usually reported for this climate and for marine conditions the corrosion degradation of infrastructure is an issue of paramount importance.

This chapter presents studies about the degradation of some engineering materials, such as austenitic stainless steels (localised corrosion in chloride containing electrolytes) and atmospheric corrosion of copper and nickel-iron alloys from both approaches experimental electrochemical tests and theoretical calculations respectively. The evaluation of the corrosion process of stainless steels in chloride containing solutions and atmospheric corrosion of copper in a marine tropical-humid climate are presented and discussed making emphasis on the electrochemical techniques used. On the other hand, a computational simulation indicated weakening of metal bonds in Fe-Ni (111) surfaces due to interaction with CO after adsorption of this compound. The union weakening observed can be associated with alloy embrittlement by the decohesion mechanism.

It is worth mentioning that one important contribution of the "Disciplinary research group: Corrosion Science and Engineering" of the Centre for Corrosion Research of the Autonomous University of Campeche, MEXICO has been the use for the very first time of the recursive plots methodology for the analysis of current and potential time series from electrochemical noise measurements for studies of localised corrosion. With such approach it was possible to assess changes in the dynamics of the degradation process and to separate the contribution of different phenomena.

Novel electrochemical techniques and advanced methods for data analysis are the base for the understanding of thermodynamic and kinetics aspects involved on the corrosion degradation of engineering materials such as copper, carbon steel and stainless steels. Electrochemical noise (EN), galvanostatic cathodic reduction (CR), scanning reference electrode technique (SRET), double loop electrochemical potentiokinetic reactivation method (DLEPR) and electrochemical impedance spectroscopy (EIS) are some of the electrochemical methods used to study the corrosion degradation process of stainless steels and other engineering metals.

The SRET has been used for the quantitative assessment of localized dissolution of AISI 304 stainless steel in natural seawater and in 3.5% NaCl solution at room temperature (25 °C) (González-Sánchez, 2002; Dzib-Pérez, 2009). Changes in the dynamics of intergranular corrosion of AISI 304 stainless steel as a function of the degree of sensitisation (DOS) was evaluated by EN using recurrence plots for the analysis of current time series (García-Ochoa et al., 2009). Also the microstructure dependant short fatigue crack propagation on AISI 316L SS was distinguished from localised corrosion taking place during corrosion fatigue tests using EN (Acuña et al., 2008).

The information presented here was divided in two main sections: Atmospheric corrosion and Localised corrosion, followed by a final section of general conclusions.

2. Atmospheric degradation of engineering alloys

2.1 Atmospheric corrosion of Cu in tropical climates

Degradation of engineering alloys due to atmospheric corrosion is the most extended type of metal damage in the world. During many years, several papers have been published in this subject; however, most of the research has been made in non-tropical countries and under outdoor conditions. Tropical climate is usual on equatorial and tropical regions and is characterized by high average temperature and relative humidity with considerable precipitation during the major part of the year. Due to these conditions a high corrosion rate of metals is usually reported for this type of climate. In coastal regions like the Gulf of Mexico (Yucatan Peninsula), there is a natural source of airborne salinity which plays an important role in determining corrosion aggressivity of these regions (Mendoza & Corvo, 2000; Cook et al., 2000). The presence of anthropogenic contaminants, particularly sulphur compounds produced at the oil and manufacture industries and transportation have also an important effect on the atmospheric corrosivity of tropical-humid regions. The atmospheric corrosion rate of metals depends mainly on the time of wetness (TOW) and concentration of pollutants; however, if the differences in the corrosion process between outdoor and indoor conditions are taken into account, the influence of direct precipitation such as rain is very important for outdoor and negligible for indoor conditions. The acceleration effect of pollutants could change depending on wetness conditions of the surface, so the influence of the rain time and quantity should be very important in determining changes in corrosion rate (Corvo et al., 2008).

Dew or humidity condensation is considered a central cause for the corrosion of metals. Its formation depends on the relative humidity (RH) and on the changes of temperature. Because dew does not wash the metallic surface, the concentration of pollutants becomes relatively high in the thin layer of electrolyte and could be much more aggressive than rain. Rain gives rise to the formation of a thick layer of water and also adds corrosive agents such as H^+ and SO_4^{2-}, however it can wash away the contaminants as well. It does depend on the intensity and duration of the rainfall.

Corvo et al., made some recommendations in order to improve the methodology for estimation of TOW-ISO which include among others the establishment of limit of air temperature ranges dividing it in two categories: from 0 °C to 25 °C and higher than 25 °C. Also the inclusion of time and amount of rain as an additional variable, taking into account the washing and cleaning effect of rain, limiting the use of TOW-ISO to outdoor and not highly contaminated environments (Corvo et al., 2008).

Together with the assessment of atmosphere corrosivity, the actual corrosion rate of metals must be evaluated in order to have the complete body of information about the phenomenon. The rate of atmospheric corrosion of metals is evaluated mainly by two procedures: the gravimetric method (mass loss measurements) and using electrochemical techniques like cathodic reduction measurements, electrochemical impedance spectroscopy, electrochemical noise and potentiodynamic polarization. Electrochemical techniques are applied to assess the corrosion rate of metals exposed to the atmosphere and to other aggressive environments because they provide instantaneous corrosion rate values and in most of the cases can be considered as non-intrusive methods. For example linear sweep voltammetry and cathodic reduction (Chronopotentiometry) are two electrochemical techniques that have been successfully used for the quantitative analysis of copper oxides formed during the atmospheric corrosion of this metal (Nakayama, 2001, 2007).

On the other hand, electrochemical noise (EN) was used as a novel approach to study atmospheric corrosion as it is able assess the protective properties of the corrosion products formed on metal surfaces. This technique involves the recording of current fluctuations taking place between two similar electrodes separated by a wetted cloth; one electrode is the surface under evaluation whilst the other is a clean non-corroded sample of the same metal. The current or potential fluctuations measured are associated with the electrochemical behaviour of the corroded metal. It can be considered as a nonintrusive technique as the metal sample is not perturbed by any electrical signal. The EN methodology has been applied successfully for the study of atmospheric corrosion in artificial and natural atmospheres giving information on the severity and morphology of the attack as a function of time (García-Ochoa et al., 2008, Torres et al., 2006).

Electrochemical Noise measurements were applied to study the protection level offered by corrosion products formed on samples of Cu exposed in different outdoor atmospheres during a relatively short periods. Also chronopotentiometric measurements (Cathodic reduction) were conducted to determine the presence of different copper compounds forming the patinas. The results from EN measurements in terms of the noise resistance (Rn) were consistent with the corrosion rate obtained from mass loss measurements and cathodic reduction as shown in figure 1, (García-Ochoa et al., 2008).

Cathodic reduction test (galvanostatic cathodic polarisation) applied on samples in an oxygen free KCl solution gave corrosion rate values very similar to those calculated from electrochemical noise and mass lose measurements.

Samples of Cu were exposed short time periods at 7 different outdoor sites. The micro-climates of these sites had remarkable effect on the atmospheric corrosion of Cu and on the kind of corrosion products formed on the surface. Electrochemical noise analysis allowed assessing the protective characteristics of the corrosion products which is related to the corrosion process and the quantity of dissolved metal.

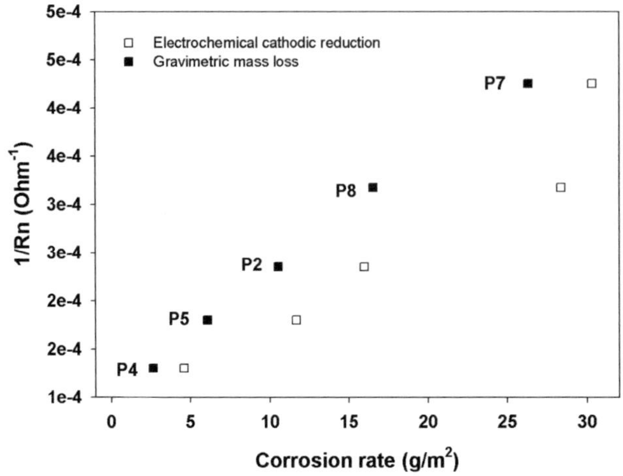

Fig. 1. The Rn values plotted against corrosion rate (gm^{-2}) obtained from mass loss and cathodic reduction tests.

The inflection points observed in the potential vs time curves gave information of the kind of compound reduced during the polarisation and the time at which the inflections appear is the time used to calculate the quantity of corroded metal as shown in Figure 2. A plateau at around -1,146 mV is clearly defined. This plateau has been associated with the presence and reduction of Cu_2S, which is a crystalline compound that has a more negative reduction potential than the copper oxides (Itoh et al., 2002, Watanabe et al., 2001). The cathodic reduction results indicated a major presence of sulphur compounds in sample 7.

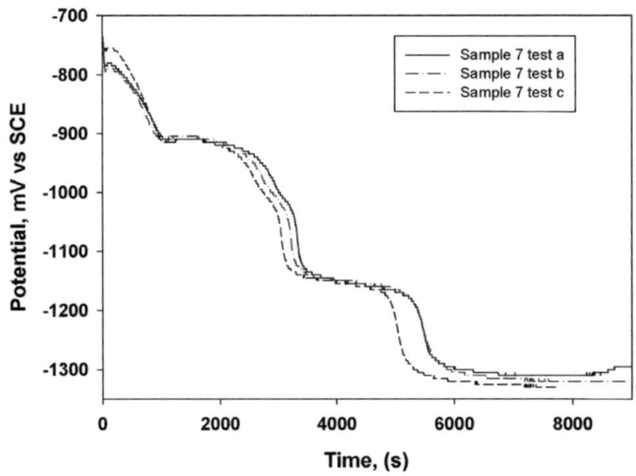

Fig. 2. Potential–time curves from galvanostatic polarisation of Cu samples in 0.1 M KCl.

The electrochemical noise technique was able to evaluate the protection level of corrosion products formed on copper samples during a relatively short exposure period to different outdoor atmospheres in terms of Rn. This parameter showed a proportional relationship with the aggressiveness of the atmosphere. The application of electrochemical noise using two electrodes offers the possibility of determining the sites where corrosion is more intense, the higher the amplitude of the signal, the higher the corrosion rate. This methodology permits evaluation using the natural surface electrolyte formed during atmospheric corrosion. The three different methods used to evaluate the atmospheric corrosion of copper: gravimetric analysis, electrochemical noise (Rn), and chronopotentiometry indicate the same pattern as a function of the exposure site. Using cathodic reduction it was possible to determine the presence of copper sulphide in the copper corrosion products, indicating the significant influence of H_2S in the atmospheric corrosion of copper.

Atmospheric degradation of engineering materials takes place also in dry conditions in which electrochemical reactions are not involved (electrochemical corrosion). This other form of degradation involves metallic surface-gas interactions. In order to get an insight of this phenomenon, theoretical and simulation studies of molecular level are carried out for different groups around the world. The case of the effect of CO adsorbed on the degradation of a Ni-Fe alloy is presented in the next sub-section.

2.2 A computational study of CO adsorption on a Ni-Fe surface

The materials used for industrial process are generally Fe- and Ni-based alloys that offer high corrosion and creep resistance. However, when the material is exposed to gases containing carbon, e.g. CO, it can pick up carbon (Grabke, 1998). The resistance to thermal cycling is reduced and there is a danger of cracks developing in the material. In consequence, the understanding of the interaction of carbon monoxide with nickel-iron alloys can be useful in order to reduce this undesirable behavior.

As an example, a combination of flow reactor studies and electron microscopy techniques has been used to investigate the way the composition of iron-nickel alloy particles influence the growth characteristics of carbon deposits formed during the decomposition of ethane at temperatures over the range 815–865 °C. Major differences in the selectivity patterns of alloys were evident with the amount of solid carbon catalytically produced being significantly higher on a Fe–Ni (5:5) powder than on a Fe–Ni (8:2) sample. Examination of the deposit revealed the existence of two types of structures, carbon nanofibers and a graphite shell-like material, both of which contained associated metal particles (Rodriguez et al., 1997). The information on literature about studies of the adsorption of CO on Fe/Ni alloys at quantum level is quite limited. This section presents a study of the CO chemisorption on a FeNi(111) surface based on calculations in the framework of the Density Functional Theory (DFT) (Hohenberg & Kohn, 1964).

The exchange-correlation potential (V_{XC}) considered within the generalized gradient approximation (GGA) proposed by Perdew et al., (Perdew et al., 1996) and the self-consistent total energy method, as implemented in the SIESTA Package code (Ordejon et al., 1996), has been used here. This methodology has been successfully applied for the study of several kinds of interactions (Sánchez-Portal et al., 2004). The electron–ion interactions are treated by means of norm conserving pseudopotentials in accordance with the Troullier-Martins procedure (Troullier & Martins, 1991). For the base set, a double zeta basis set plus polarization functions (DZP) was used. The atomic orbitals were slightly excited (0.01 Ry) in order to limit the range of the pseudo-atomic base orbitals (Sankey & Niklewski, 1989).

A uniform grid in real space with a mesh cutoff of 450 Ry was used for calculations. The Brillouin zone was sampled, and total energy converged correspondingly to the number of k points resulting in the Monkhorst-Pack matrix diagonal (7×7×1) (Monkhorst & Pack 1976).

To understand the interactions between the atoms, we used the concept of COOP (Crystal Orbital Overlap Population) curves. A COOP curve is a plot of the overlap population weighted DOS (density of states) vs. energy. The integration of the COOP curve up to the Fermi level (E_f) gives the total overlap population of the bond specified and it is a measure of the bond strength.

The interaction between the CO molecule and the FeNi(111) surface was studied using a two dimensional slab of finite thickness in order to have the best simulation of the semi-infinite nature of the metallic surface. A three-layer slab was employed as a compromise between computational economy and reasonable accuracy. The FeNi(111) surface was represented by a 108 atoms (50:50) distributed in three layers (FCC arrangement). It was found the final configuration of the CO/FeNi(111) system using the Spanish Initiative for Electronic Simulations with Thousands of Atoms (SIESTA) method (Soler et al. 2002). A geometry optimization was performed applying relaxation calculations. The top two layers of the substrate were allowed to relax together with the adsorbate while the bottom layer was kept fixed in the bulk position. Table 1 presents the C-surface distances and the relative minimum energy corresponding to the CO location for each adsorption site by SIESTA calculations.

Adsorption Site	C-surface distance (Å)	Relative Energy (eV)
1	1.79	0.58 (L)
2	1.83	0.41(L)
3	1.40	0.19 (NL)
4	1.20	0.00 (NL)
5	1.43	0.278 (L)
6	1.38	0.232 (L)
7	1.38	0.069 (NL)

L: local minimum energy

NL: no local minimum energy

Table 1. Carbon-surface distances calculations and, carbon-surface distance, relative energy and type of the minimum energy position by SIESTA calculations for the CO adsorption sites on the FeNi(111) surface.

The 1, 2, 5 and 6 sites present a local minimum energy showing the molecule optimum localization at these sites. On the other hand, the 3, 4 and 7 sites are not stable and correspond to transition states, since CO relaxes to other sites when initially put at these ones. Finally, the most stable location for CO on the FeNi(111) surface corresponds to an intermediate position between 1 and 4 sites, where the C atom is positioned in the middle of two Ni atoms and a neighbouring Fe atom. A schematic view of the CO location is shown in Figure 3 (bottom).

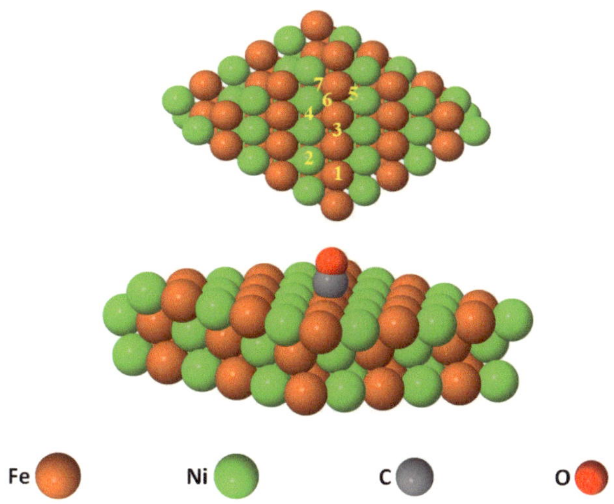

Fig. 3. Schematic (a) initial sites for CO adsorption and (b) final view of the CO location, on the FeNi(111) surface

For the final configuration a C-O distance of 1.20 Å and a C-surface distance of 1.35 Å were found which are in agreement with results reported for the adsorption of CO on both Fe and Ni single crystal surfaces, respectively (Jiang & Carter, 2004; Karmazyn et al., 2003; GajdoŠ et al., 2004; Peters et al., 2001). For the electronic structure calculations, the density of states (DOS) and the crystal orbital overlap population (COOP) curves for the CO/FeNi(111) system were determined in order to analyze the adsorbate-surface interactions. Figure 4 (middle) shows the DOS plots for the CO/FeNi(111) system.

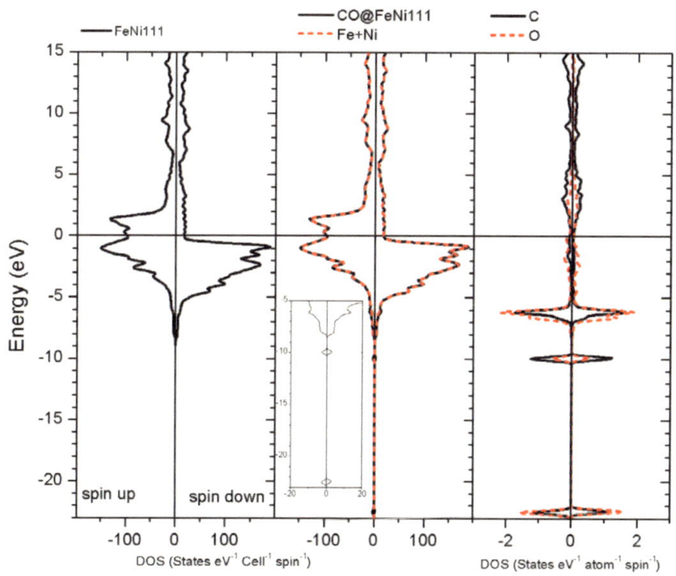

Fig. 4. Total DOS for the CO/FeNi(111) system (middle), total DOS for the clean FeNi(111) surface (left) and projected DOS for CO on the FeNi(111) surface (right).

The small contribution of the CO to total DOS is due to its low concentration. For a major view, Figure 4 (right) presents a plot of CO states projection after adsorption. To understand the interactions between the atoms, concept of COOP (crystal orbital overlap population) curves was used. The atomic orbital occupation and the OP values for the atoms that participate in the absorbate-substrate interactions were also calculated (see Table II).

Atom	Orbital occupation			Charge	Bond	Distance (Å)	OP
	s	p	d				
Fe_{nn}	0.765	0.622	6.537	0.076 [a]	Fe_{nn}-C	2.09	0.185 [a]
	0.848	0.591	6.477	0.084 [b]	Fe_{nn}-NN	------	0.803 [a] 0.941 [b]
Ni_{nn}	0.788	0.856	8.503	-0.147 [a]	Ni_{I}-C	1.94	0.246 [a]
	0.871	0.735	8.494	-0.100 [b]	Ni_{I}-NN	------	0. 845 [a] 0.989 [b]

[a] In the CO@Feni(111)

[b] In the FeNi(111)

nn: nearest neighbor to C

NN: up to 3rd nearest neighbor in the metallic surface

Table 2. Atomic orbital occupations and net charges for the CO neighboring Fe and Ni atoms, and the corresponding OP values and distances by SIESTA calculations.

The C (of the CO molecule) bonds with nearest neighbors Ni and Fe surface atoms reported Ni-C (1.94 Å) and Fe-C (2.09 Å) OP values of 0.246 and 0.185, respectively. Compare these new interactions with the metal-metal interaction (isolated FeNi matrix), the Ni-C and Fe-C interactions have OP values that correspond to 25 % and 20 % of the metallic bond OP, respectively. The COOP curves for the new interactions correspond to mainly bonding interactions as can be seen in Figure 5.

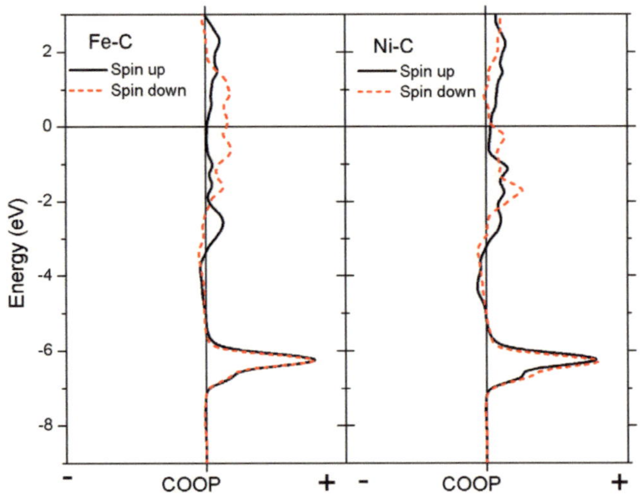

Fig. 5. COOP curves for Fe-C and Ni-C interactions.

The substrate-adsorbate interactions mainly involve the s and p orbitals of Ni whose populations decrease 9.53 % and increase 16.46 %, respectively compared with a clean surface. The Ni d orbital populations only decrease to about 0.11 %. On the other hand, the Fe orbital occupations are modified and the major changes are also noticed in the s and p atomic orbitals whose populations decrease 9.78 % and increase 5.24 % respectively, after CO adsorption. The Fe d populations only decrease to about 0.93 %. In general, there is observed an electron transfer from CO to the Fe and Ni nearest neighbors, then the surface-layer of the slab is negatively charged relative to the bulk due to CO interaction (see Table II). A large bonding OP between C and both Ni and Fe atoms appears, while the Ni-Ni, Fe-Fe and Ni-Fe OP decrease. After CO adsorption, the strength of the Ni-NN and Fe-NN bonds (NN: metallic atoms up to 3rd nearest neighbor) decreases to about 15 %. A detrimental effect on the metal bonds is observed after CO adsorption on the FeNi(111) surface and can be associated with the alloys embrittlement by decohesion mechanism.

3. Localised corrosion

3.1 Localised corrosion of stainless steels

Stainless steels are basically iron–chromium–nickel alloys, containing between 18 and 30 wt% chromium, 8–20 wt% nickel and 0.03–0.1 wt% carbon. According to metallurgical structure stainless steels are divided into three groups: austenitic "γ" face centred cubic (fcc), ferritic "α" body centred cubic (bcc), and martensitic (body centred tetragonal or cubic). There is another stainless alloy, duplex ("γ - α"), which possesses a two-phase microstructure with approximately equal amounts of austenite and ferrite (Marshall, 1984).

Austenitic stainless steel of the series 300 such as the UNS 30400 (S30400 SS) is used in a wide range of applications due to its acceptable corrosion resistance in non-chloride containing environments and good weldability. This kind of steel loses its corrosion resistance when is cooled slowly from the solution anneal temperature around 1273 K (1000 °C) or is reheated in the range from 823 K (550 °C) to 1123 K (850 °C). In this temperature range there is a tendency to precipitate chromium-rich carbides as the alloy enters the carbide plus austenite phase field (Marshall, 1984; Lacombe et al., 1993). Precipitation of carbides such as $M_{23}C_6$ and M_7C_3 occurs primarily at the austenite grain boundaries which are heterogeneous nucleation sites. The chemical composition in the vicinity of the grain boundaries can be altered by the precipitation of the chromium rich particles (Lacombe et al., 1993). This phenomenon is called sensitization and prompts the resulting chromium-depleted zones at the grain boundaries to be susceptible to intergranular corrosion (Terada et al., 2006). This is a well known form of localised corrosion on stainless steels in particular on sensitised austenitic grades and look like the examples presented in Figure 6.

There are several published works about the corrosion of stainless steels which deal with different metal-electrolyte systems using different electrochemical techniques (Burstein, 2004; Curiel, 2011; Isaacs, 1989; Newman, 2001; Sudesh, 2007; Turnbull, 2006). Localised attack in the form of crevice and pitting corrosion is the most insidious and common initiation stage for the development of cracks under static or cyclic mechanical loading (Zhou, 1999; Akid, 2006; González-Sánchez, 2002; Acuña, 2005).

The effect of sensitisation on the corrosion resistance of stainless steels is difficult to determine quantitatively using conventional polarisation electrochemical methods, owing to the negligible weight loss involved. The microscopic dimension of the chromium depleted zone next to the grain boundaries is overshadowed by the unaffected bulk of the grains in many conventional corrosion tests.

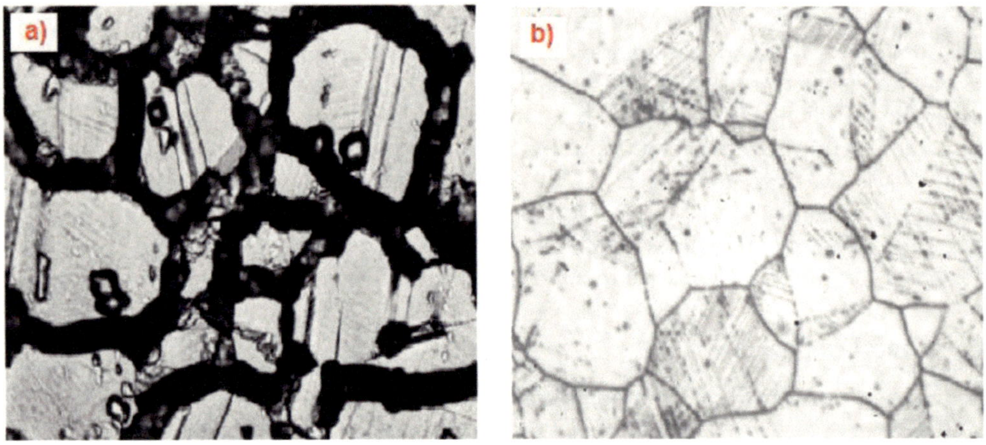

Fig. 6. Micrograph of sensitised austenitic stainless steel, a) sample of AISI 304 steel after high anodic polarisation in artificial seawater and b) sample after cyclic polarisation in H_2SO_4 + KSCN solution

Several electrochemical and non-electrochemical methods have been proposed and used to evaluate the degree of sensitisation of stainless steels. One of the most utilised is the electrochemical potentiokinetic reactivation test (EPR) based on Číhal's method (Číhal & Štefec, 2001).

Due to its quantitative nature and reproducibility, this method has been standardized by ASTM to estimate the sensitization grade of AISI type 304 and 304L stainless steels (ASTM, 1994).

Other research group proposed the double loop electrochemical potentiokinetic reactivation method (DLEPR) for determining the sensitization grade of stainless steels (Majidi et al., 1984). This author compared the results of the new method, the single loop and the acid test, and observed a wood agreement between measurement made with double loop and single loop EPR test giving a quantitative measure of sensitization based on the ratio of active peak currents on the forward and reverse scans Ir/Ia when the polarization is carried out in a 0.5 M H_2SO_4 + 0.01 M KSCN solution and a scan rate of 100 mV min-1.

Figure 7 presents the DLEPR curves of AISI 304 stainless steel for the case of samples taken from the heat affected zone (HAZ) after Gas metal arc welding (GMAW) and samples solution annealing at 1050 °C (red curve), (Curiel et al., 2011). The results obtained the DLEPR tests indicated clearly the effect of metallurgical condition of the stainless steel on its resistance to localised corrosion (intergranular). Samples of the HAZ presented a clear reactivation current density whereas the samples solution annealed showed just a negligible value of Ia. After the annealing treatment for 1 hr, not all Cr was dissolved completely, that is why there exists a small reactivation but several times lower than that for sensitised samples.

Electrochemical methods used to determine the sensitization intensity on stainless steels enjoyed wide expansion over the last 40 years, and the DL-EPR test has become one of the most successful due to its quantitative nature and because can be considered as a non-destructive method.

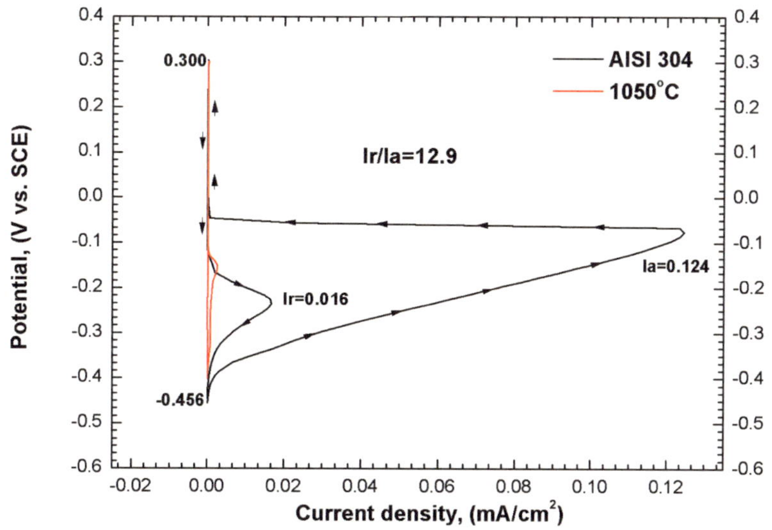

Fig. 7. DLEPR curves obtained from 304 SS samples at the HAZ after GMAW (black curve) and after solution annealing (red curve).

However as with any testing technique, attention must be paid in interpreting the results as measurements are sensitive to local changes in composition and microstructure of the alloy under study. In this sense, the use of alternative and reliable electrochemical technique has become o necessity in order to find the best way to determine the electrochemical behaviour of metallic materials or their corrosion resistance in diverse electrolytes.

During electrochemical reactions such as the corrosion of metals, electrochemical micro-cells form on the surface of the metal in contact with an electrolyte which induce potential and current fluctuations. These electrochemical fluctuations are known as Electrochemical Noise (EN) and definitively contain information about faradaic processes taking place on the electrified interface formed by a metal in contact with an electrolyte. In the second half of the 20th century there were numerous reports about the existence of these fluctuations (Iverson, 1968; Searson & Dawson, 1988). Several parameters are usually acquired from EN measurements which depend upon the method used for the data analysis (Gouveia-Caridade et al., 2004; Zaveri et al., 2007; Cottis & Turgoose, 1999). The results can be plotted as potential and current time series as the example shown in Figure 8.

Since the 90ties a preponderant importance has been given to the study of this phenomenon to explain the dynamics and mechanisms of the electrochemical processes taking place in the electrified interface (Gouveia-Caridade et al., 2004; Kearns et al., 1996; García-Ochoa et al., 1996; Hladky & Dawson, 1980; Loto & Cottis, 1987).

A number of procedures have been proposed and applied for the analysis of electrochemical noise data, from simple statistics analysis up to strategies that consider that pitting corrosion has a chaotic nature and apply non-linear methods to obtain parameters like the Lyapunov exponent (García et al., 2003).

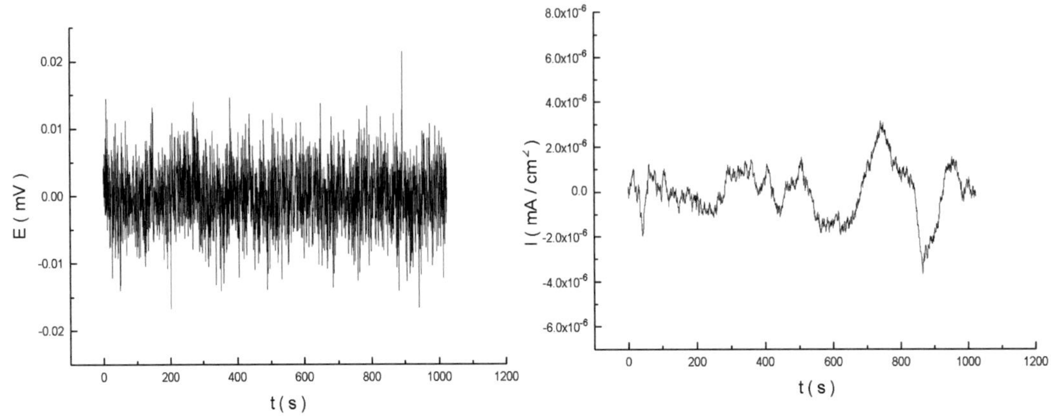

Fig. 8. Potential and current density time series obtained from electrochemical noise measurements.

Pitting corrosion studies conducted in the past using electrochemical noise measurements considered that high amplitude oscillations associated to breakdown - recovery of passive layers had a totally stochastic nature. Succeeding studies demonstrated that such approach was not correct because oscillations observed during the pitting of iron have a non-linear chaotic nature, this means processes of complex dynamics that are sensitive to initial conditions (García et al., 2003; González et al., 1997; Sazou & Pagitsas, 2003.

Visual recurrence analysis is another procedure to study the behaviour of nonlinear dynamical systems such as localized corrosion processes. This procedure has been used to differentiate between stochastic and chaotic variability. The principal instruments of the recurrence analysis are the Recurrence plots (RPs) which are especially useful for the graphical representation of multidimensional dynamic systems (Eckmann et al., 1987; Casdagli, 1998; Trulla, 1996). Recurrence plots (RPs) are a valuable tool for measure the geometry of the dynamics exploiting non-linear dependencies even in non-stationary time-series. These plots disclose distance relationships between points on a dynamical system providing a faithful representation of the time dependencies (correlations) contained in the data (Acuña et al., 2008). This is a graphical tool for the diagnosis of drift and hidden periodicities in the time evolution of dynamical systems, which are unnoticeable otherwise. Recurrence plots (RPs) are graphical tools elaborated by Eckmann et al. based on Phase Space Reconstruction (Eckmann et al., 1987). The method of RPs was introduced to visualize the time dependent behavior of the dynamics of systems, which can be pictured as a trajectory in the phase space as presented in Figure 9, (Zbilut & Webber, 1992; McGuire et al., 1997; Marwan et al., 2007).

This methodology was used also to study the dynamics of intergranular corrosion in austenitic stainless steel with different degree of sensitization (García-Ochoa et al., 2009).

The analysis of electrochemical noise in current using recurrence plots proved to be an excellent tool to evaluate the changes in the dynamics of the intergranular corrosion of AISI 304 austenitic stainless steel with different degree of sensitization. The RP´s showed that

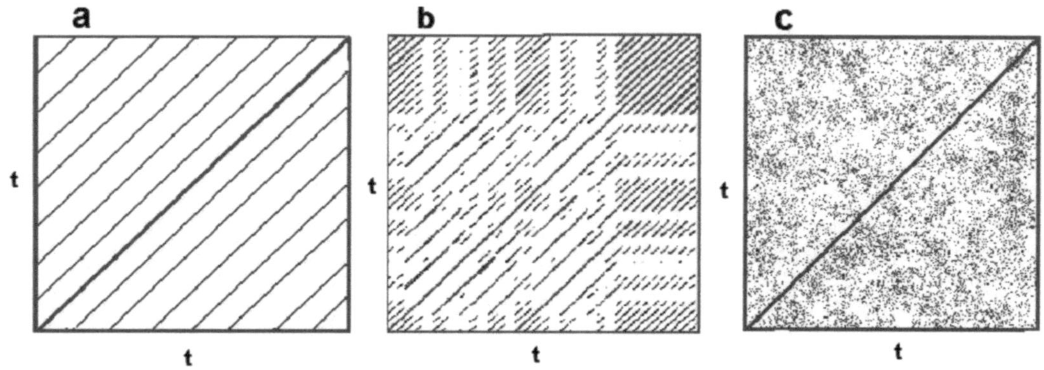

Fig. 9. Recurrence plots of (a) a periodic motion with one frequency, (b) the chaotic Rossler system and (c) of uniformly distributed noise (Marwan et al., 2007).

sensitisation causes a localised corrosion process with spatiotemporal well defined electrochemical cells interacting in the form of a dissolution process with periodic dynamics. The periodicity was determined by the increment of %D (percentage of determinism) and R% (the percentage of recurrence) as a function of the degree of sensitisation (DOS) (García-Ochoa et al., 2009).

Electrochemical noise measurements conducted during environmental assisted cracking of austenitic stainless steel detected changes in the electrochemical fluctuations which were associated to the cracking processes (Acuña, 2005; González et al., 1997). The methods applied to date for the analysis of electrochemical noise data to study the initial stages of corrosion fatigue damage (CFD) have been unable to separate the contribution due to localized corrosion from that due to crack nucleation and growth. Crack nucleation and growth involve generation of fresh active metal surfaces which interact with the electrolyte and induce changes in the amplitude of current fluctuations. This contribution to the noise signal must be different from that associated to the localised corrosion process which in principle should have dissimilar nature.

It was possible to separate the contribution of pitting corrosion for which electrochemical noise in current presented a percentage of determinism (%D = 80) higher than that associated to CF crack initiation which presented a stochastic behaviour with low %D (around 5%). This separation was possible by the use of the recurrence quantitative analysis parameter (RQA) selected: the percentage of determinism %D. Recursive Plots RPs applied to the analysis of electrochemical current noise measured during CF tests and their assessment by (RQA) represents a powerful non-linear analysis tool as it allowed us to establish clearly the dynamics of the early stages of CF cracking (Acuña et al., 2008).

3.2 Studies of localised corrosion using the Scanning Reference Electrode Technique (SRET)

The dissolution of metals during localised corrosion takes place at permanently separated sites from the bigger cathodic areas. This gives the possibility of direct measurements of the cathodic and anodic reactions through *in situ* non-intrusive studies. In order to study the kinetics of localised corrosion in its various forms it is necessary to use electrochemical techniques capable to measure variations in electrochemical activity directly on the site undergoing localised attack at the metal surface.

Measurements of the physical separation of anodic and cathodic areas, the currents flowing between them as well as the mapping of potentials in electrolytic solutions have been successfully used for the study of the processes of localised corrosion of different systems (Isaacs & Vyas, 1981; Tuck, 1983; Bates et al., 1989; Sargeant et al., 1989; Trethewey et al., 1994; Trethewey et al., 1996; McMurra et al., 1996; McMurray & Worsley, 1996). A schematic drawing of a local corrosion cell is shown in figure 10.

With the aim of determining the velocity of metal dissolution directly in active pits during pitting corrosion, Rosenfeld and Danilov (Rosenfeld & Danilov, 1967), designed an apparatus to measure the field strength in the electrolyte directly above an active pit. They employed a twin probe method by using two reference electrodes, which makes it possible to measure the potential difference ΔE in any direction between two points in the electrolyte with the aid of two non-polarisable electrodes, for example calomel electrodes.

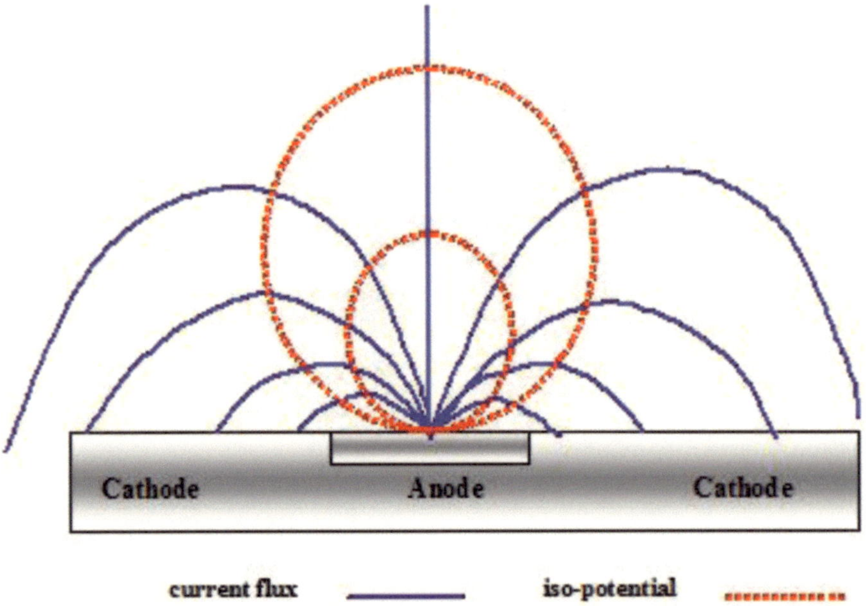

Fig. 10. Schematic of current and potential distribution in solution during localised corrosion

The equipment used for the measurements of the potential difference ΔE (ohmic potential gradients) is called the Scanning Reference Electrode (SRET). With the measurement of the electric field strength in the electrolyte over the pits it was possible to determine the current flowing from the anode points, based on the fact that, the vector of the normal component of the current density at a pre-determined point (i') in a uniform field is equal to the product of the electric field strength E and the specific conductivity of the medium κ.

The resolution of the SRET depends upon the proximity of the scanning probes to the corroding sites and the magnitude of the corrosion currents from each site. As shall be shown later, the distance between the probe and specimen surface and the conductivity of the solution governs the sensitivity of the technique. It has been reported the capability of the SRET to identify the position of localised activity in the metal surface however did not report any assessment of pit size or shape from the performed SRET measurements (Bates et al., 1989; Sargeant et al., 1989; Trethewey et al., 1994).

Tuck, (Tuck, 1983) evaluated the usefulness of reference microelectrodes in identifying local anodic and cathodic sites on aluminium alloys as they were scanned mechanically over a polished surface. The electrolyte in which the specimens were immersed for the study was shown to have a critical effect on the detectability of sites undergoing localised activity. He demonstrated that a solution of low conductivity is indispensable if the electrodes were microscopic in size. This is obviously a limitation in the application of microelectrodes because they cannot be used for real systems undergoing localised corrosion in electrolytes of high conductivity, e.g. seawater. Trethewey et al. (Trethewey et al., 1993) obtained pit life history at specific points in a 304 stainless steel specimen immersed in seawater. They showed that the measured current density adjacent to an active pit exceeded 300 times that obtained from a conventional pitting scan which was a maximum current density of 0.8 mA/cm^2. This author showed the advantages that in theory should give the use of a differential probe configuration over the conventional single ended system (Trethewey et al., 1994). The same authors indicated that with the use of SRET it is possible to study pit initiation and development, surface coating behaviour, inhibitor performance, battery performance, corrosion under hydrodynamic conditions as well as microbiological induced corrosion and stress corrosion cracking.

The SRET operation principle is based on the fact that during the localised corrosion of metals the electronic charge generated by the dissolution reaction flows from the localised anode to the cathodic sites through the metal. The high electronic conductivity of the metal induces a negligible ohmic potential difference in the metal, thus the surface of the corroding metal can be considered as a plane of constant potential. However within the aqueous electrolyte in contact with the corroding metal the ionic flow that develops to complete the corrosion cell produces ohmic potential gradients owing to the low electric conductivity of the electrolyte. As shown in figure 7, these potential gradients may be described as a series of iso-potential lines lying in perpendicular direction to the lines of ionic current flux. The activity can be assessed in terms of the current emanating from the sites undergoing local dissolution. By scanning a non-polarizable reference probe containing a fine capillary tip parallel and very close to the metal surface, the ohmic potential gradients generated in the electrolyte by localised anodic currents can be measured. It must be emphasised that the SRET does not directly measure the potential variations in the surface of the metal, but it responds to the ohmic potential gradients originated by ionic fluxes in the solution. SRET is a powerful equipment allowing real-time localised electrochemical activity to be managed and fully quantified.

From a study of pitting corrosion of austenitic stainless steel in artificial seawater SRET map scans were obtained which gave information about the electrochemical activity emanating from sites undergoing localised dissolution. A commercial SRET (Uniscan instruments model SR100) in which a cylindrical specimen (sealed tube or bar) is the working electrode, immersed in the electrolyte and rotated at precise speeds in the range 5–250 rpm was used.

Figure 11 presents a schematic of the SRET equipment used for the study.

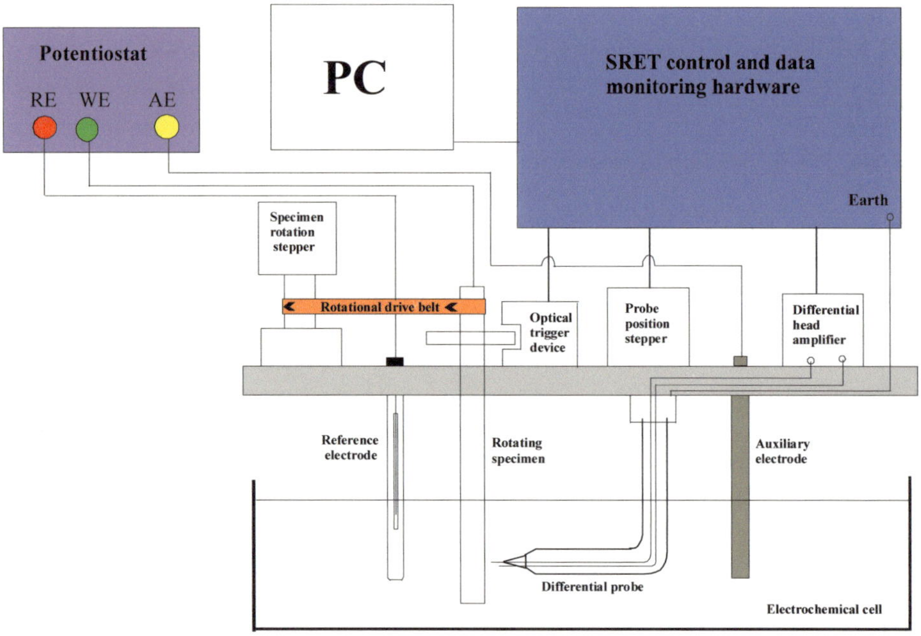

Fig. 11. Schematic representation of the SRET SR100.

A calibration procedure is mandatory and important aspect in order to assess the anodic current density associated with the ohmic potential gradient measured by the SRET. In this case a punctual source of current (a gold wire of 200 μm diameter) is immersed in the solution and polarised galvanostatically. The SRET measures a potential gradient which is then associated to the applied anodic current. A line scan and a map scan of the punctual source of current is presented in Figure 12.

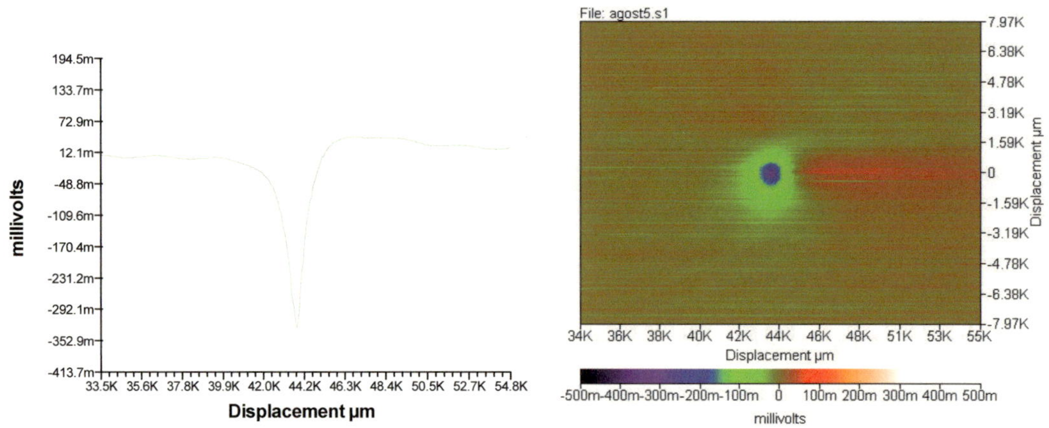

Fig. 12. Line and area map scans for a punctual current source in artificial seawater with an applied current density of i = 33.8 mA/cm^2, Maximum output signal = 0.327 mV.

Calibration experiments for the SRET showed that as the electrolyte conductivity increased, higher applied currents were needed in order to induce a signal detectable by the equipment. In low conductivity electrolytes (0.014 mS/cm), the application of a current of 6.42 nA to a Punctual Current Source (PCS) was sufficient to generate a detectable signal.

However, in electrolytes with a conductivity of 55.8 mS/cm (3.5 wt% NaCl), the SRET detected a minimal output signal generated by applied currents higher than 3.73 mA as shown in Figure 13, (Dzib-Pérez et al., 2009).

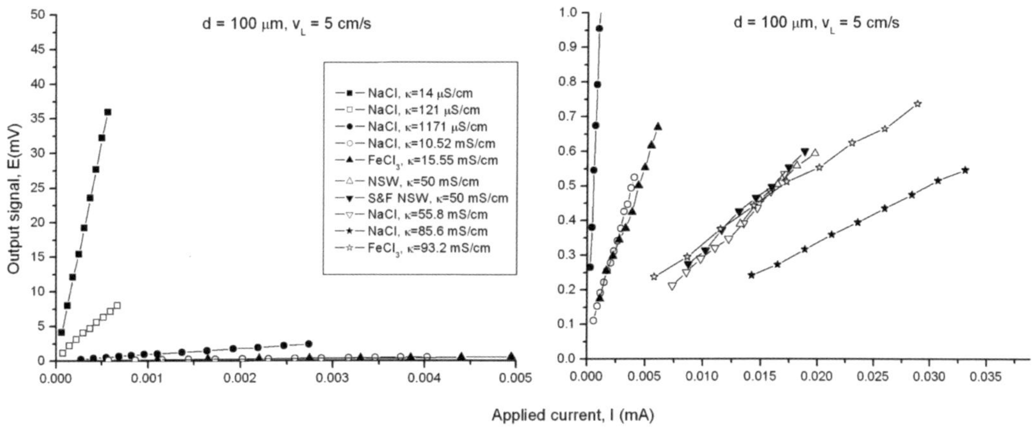

Fig. 13. Output signal vs applied current in electrolytes with different conductivity, 100 μm of separation surface-probe tip and a rotation rate of 5 cm/s. Magnification at the right to observe the response in high conductivity electrolytes.

The rotation rate of the working electrode has significant effect on the output signal for measurements in low conductivity electrolytes. The higher the rotation rate, the higher the slope of maximum signal detected vs applied current, which means better responsiveness (higher sensitivity of SRET) as shown in Figure 14.

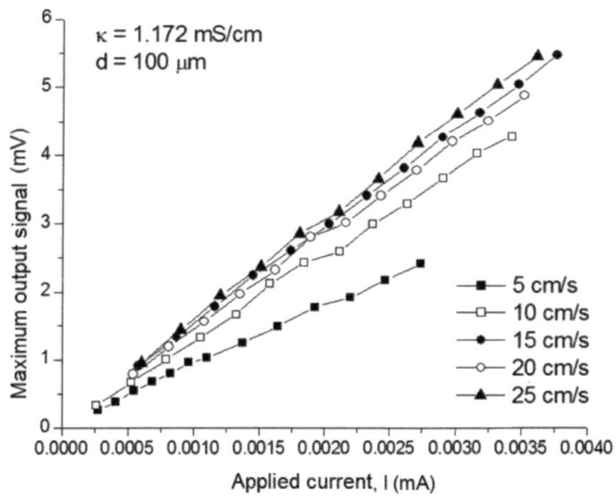

Fig. 14. Output signal vs applied current in NaCl solution (k = 1.172 mS/cm), with a separation probe tip to PCS of 100 μm and five different rotation rates.

Nevertheless, the resolution (WHM) of the SRET instrumentation decreases as the rotation rate increases. For the case of studies of localised corrosion in diluted electrolytes, this is a useful finding. It is of paramount importance an accurate understanding of the effects that operating parameters of the SRET equipment has on the measured output signal (Dzib-Pérez et al., 2009).

Studies of pitting corrosion of AISI 304 stainless steel in natural seawater were conducted performing SRET measurements on samples under cyclic anodic potentiodynamic polarisation from the open circuit potential to the repassivation potential as shown in Figure 15, (Dzib-Pérez, 2009; González-Sánchez, 2011). This study allowed quantitatively assessing the dissolution rate of stable pits and monitoring from initiation to repassivation (Dzib-Pérez, 2009).

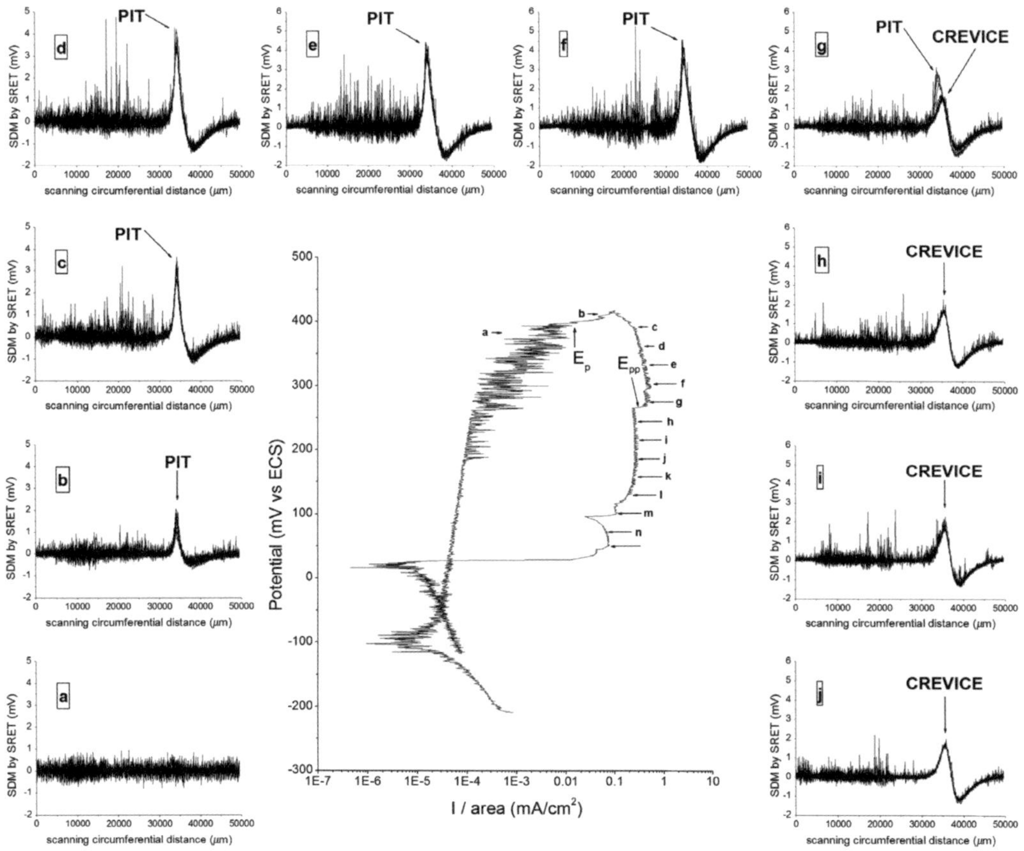

Fig. 15. SRET measurements conducted during potentiodynamic polarisation of AISI 304 stainless steel in natural seawater.

For a number of active pits generated on the AISI 304 stainless steel, measurements of the dissolved volume of metal per pit was determined by two methods. One was from material removal considering that pits had a spherical and elliptical geometry (Turnbull et al., 2006; González-Sánchez, 2002), and the second was done using SRET measurements of current density (Dzib-Pérez, 2009; González-Sánchez, 2002). By summing the SRET measured current density of the pit as a function of time and using Faraday´s law the quantity of dissolved metal was determined. The volume of dissolved metal per pit was calculated using the density of the AISI 304 stainless steel. The results showed acceptable agreement as can be seen in Figure 16.

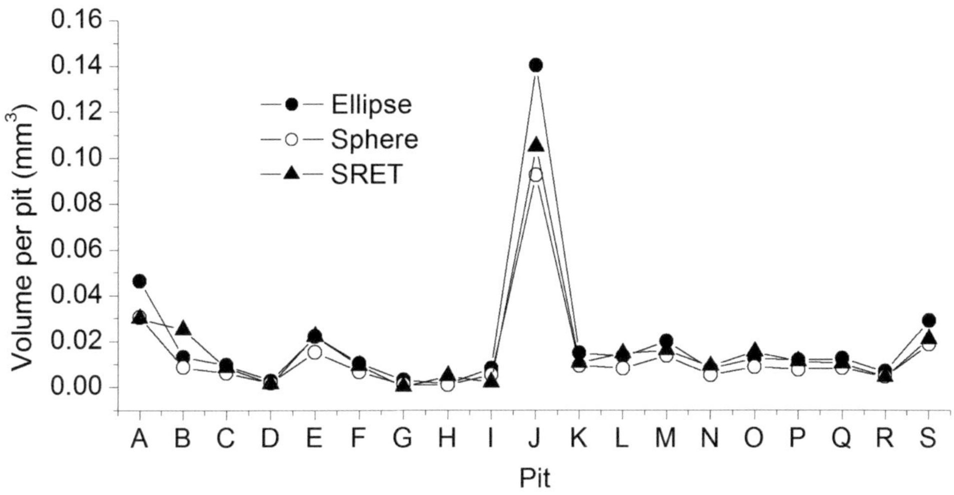

Fig. 16. Volume of dissolved metal per pit of AISI 304 stainless steel in natural seawater determined by material removal and SRET measurements.

Besides the capability of this technique to determine the position of the active pits, it is able to assess semi-quantitatively the dissolution rate in terms of localised current density. Calculations of the pit depth from the values of current density obtained from SRET agree well with the physical pit depth determined by material removal. SRET measurements showed that under potentiostatic control, active pit growth on 304 SS stainless steel in natural seawater takes place with an increase of localised current with time.

4. Conclusions

The degradation of engineering materials is a topic of paramount importance due to diverse forms in which the phenomenon takes place. From the apparently benign (non-aggressive) environment created by the atmosphere to really demanding media such as marine conditions and sour gas and oil exploitation, conduction and processing, metals and alloys will face some kind of degradation. Also at high temperatures, chemical reactions take place producing metal degradation on dry conditions.

Electrochemical novel techniques as well as novel analysis methods are currently used in order to get an insight of the mechanisms of metals degradation in aqueous environments. Sensitive techniques like electrochemical noise are able to give information of changes in the dynamics of electrode reactions that take place during metallic corrosion. SRET measurements provide the possibility of quantitative, or at least semi-quantitative assessment of the localised dissolution rate in terms of current density.

On the other hand, the understanding of chemisorption of small gas molecules on transition metal surfaces is crucial to obtain a molecular level understanding of the mechanism of heterogeneous catalysis. From simulations at molecular level it was observed a metal bond weakening of 15% after CO adsorption. A detrimental effect on the metal bonds is observed after CO adsorption on the FeNi(111) surface and can be associated with the alloys embrittlement by decohesion mechanism.

5. Acknowledgment

The authors would like to thank The Autonomous University of Campeche, MEXICO for the support given to the Centre for Corrosion Research in order to carry out the studies presented in this chapter. Also we would like to thank undergraduate and graduate students who participate in all the studies/projects about corrosion, and finally but not less important is to thank The National Council of Science and Technology (CONACYT) for the financial support given to the authors to conduct several research projects whose results were also presented in the chapter.

6. References

Acuña, N. (2005). "Fatigue Corrosion Cracking of an Austenitic Stainless Steel using Electrochemical Noise Technique", *Anti. Corr. Meth. Mater.* Vol. 52, pp. 139-144.

Acuña, N., García-Ochoa, E., & González-Sánchez, J. (2008). Assessment of the dynamics of corrosion fatigue crack initiation applying recurrence plots to the analysis of electrochemical noise data *Int J Fatigue* Vol. 30, pp. 1211-1219

Akid, R., Dmytrakh, I. & Gonzalez-Sanchez, J. (2006). "Fatigue Damage Accumulation: The role of corrosion on the early stages of crack development", *Corros. Engng. Sci. & Techno*, Vol. 41, pp. 328-335.

ASTM Standards 1994, Standard G 108-94 In: Annual Book of, American Society for Testing and Materials, Philadelphia,PA, 1994, p. 444.

Bates, S., Gosden, S., & Sargeant, D. (1989). "Design and development of Scanning Reference Electrode Technique for investigation of pitting corrosion in FV 448 turbine disc steel", *Materials Science and Technology*, Vol. 5, pp. 356

Burstein, G., Liu, C., Souto, R., & Vines S. (2004). Origins of Pitting Corrosion, *Corros. Engng, Sci. Tech.* Vol. 39, pp. 25–32.

Casdagli, M. (1997). *Physica D* Vol. 108: 12

Čihal, V., & Štefec, R. (2001). *Electrochim Acta* Vol. 46, pp. 3867

Cook, D., Van Orden, A., Reyes, J., Oh, S., Balasubramanian, R., Carpio, J. & Townsend, H.(2000). Atmospheric corrosion in marine environments along the Gulf of Mexico, in: S.W. Dean, G. Hernández Duque-Delgadillo, J. Bushman (Eds.), Marine Corrosion in Tropical Environments, ASTM STP 1399, American Society for Testing and Materials, West Conshohocken, PA.

Corvo, F., Pérez, T., Martin, Y., Reyes, J., Dzib, L., González-Sánchez, J., & Castañeda, A. (2008). "Time of Wetness in tropical climate: Considerations on the estimation of TOW according to ISO 9223 standard", *Corros. Sci.*, Vol. 50, pp. 206-219.

Cottis, R., & Turgoose, S. (1999). Electrochemical impedance and noise. NACE, Houston, TX

Curiel, F., García, R., López, V., & González-Sánchez J. (2011). "Effect of magnetic field applied during gas metal arc welding on the resistance to localised corrosion of the heat affected zone in AISI 304 stainless steel, *Corros. Sci.*, Vol. 53, pp. 2393-2399.

Dzib-Pérez, L., González-Sánchez, J., Malo, J., & Rodríguez F. (2009). The effect of test conditions on the sensitivity and resolution of SRET signal response, *Anticorrosion Methods and Materials*, Vol. 56, No. issue 1, pp. 18-27

Dzib-Pérez, L., Ph.D. thesis, National Autonomous University of Mexico, Mexico, (2009).

Eckmann, J., Kamphorst, S., & Ruelle, D. (1987). Recurrence plot of dynamical system. *Europhys Lett* Vol. 4:973–7.

GajdoŠ, M., Eichler, A., & Hafner, J. (2004). CO adsorption of close-packed transition and Ninoble metal surfaces: trends from ab initio calculations. *J. Phys.: Condens. Matter.* Vol. 16, pp. 1141

García-Ochoa, E., González-Sánchez, J., Acuña, N., & Euan, J. (2009). Analysis of the dynamics of Intergranular corrosion process of sensitised 304 stainless steel using recurrence plots, *J. Applied Electrochem*, Vol. 39, pp. 637-645.

García-Ochoa, E., Ramírez, R., Torres, V., Rodríguez, F., & Genesca J. (2002). *Corrosion* Vol. 58, pp. 756.

García-Ochoa, E., González-Sánchez, J., Corvo, F., Usagawa, Z., Dzib-Pérez, L., & Castañeda, A. (2008). Application of electrochemical noise to evaluate outdoor atmospheric corrosion of copper after relatively short exposure periods, *J. Appl. Electrochem.*, Vol. 8 pp. 1363-1368.

García, E., Uruchurtu, J., & Genescá, J. (1996). *Afinidad* Vol. 53, pp. 215

García, E., Hernández, M., Rodríguez, F., Genescá, J., & Boerio, J. (2003). Oscilation and chaos in pitting corrosion of steel. *Corrosion* Vol. 59:50–8.

González, J., Salinas, V., García, E. & Díaz, A., (1997). Use Electrochemical Noise to detect the initation and propagation of stress corrosion cracks in 17-4 PH steel. *Corrosion* Vol. 53, pp. 693–699.

González-Sánchez, J. (2002). Corrosion fatigue initiation in stainless steels: The scanning reference electrode technique Ph.D. thesis, Sheffield Hallam University, U.K

González, J., Dzib, L., & Malo, J., Evaluation of pit dissolution rate on AISI 304 stainless steel with scanning reference electrode technique, (On preparation)

Grabke, H. (1998). Carburization: A High Temperature Corrosion Phenomenon, Elsevier, Amsterdam, The Netherlands.

Gouveia-Caridade, C., Pereira, I., & Brett, M., (2004). *Electrochim Acta* Vol. 49, pp. 785
Hladky, K., & Dawson, J. (1980). *Corros. Sci.* Vol. 21, pp. 317.
Hohenberg, P., & Kohn, W. (1964). *Phys. Rev.* Vol. 136, pp. B864–B871.
Isaacs H., The localised Breakdown and Repair of Passive Surfaces during Pitting, *Corros. Sci.* Vol. 29, pp. 313 - 323.
Isaacs, H., & Vyas, B. (1981). "Scanning Reference Electrode Techniques in Localised Corrosion", *Electrochemical Corrosion testing*, ASTM STP 727, p. 3, F. Mansfeld and U. Bertocci eds.
Itoh, J., Sasaki, T. & Ohtsuka, T. (2000). *Corros Sci* Vol. 42, pp. 1539
Iverson, W. (1968). *J. Electrochem. Soc.* Vol. 115, pp. 617
Jiang, D., Carter, E. (2004). Adsorption and dissociation of CO on Fe(110) from firstprinciples, *Surface Science* Vol. 570, pp. 167–177
Karmazyn, A., Fiorin, V., Jenkins, S., & King, D. (2003). First-principles theory and microcalorimetry of CO adsorption on the {211} surfaces of Pt and Ni, *Surface Science* Vol. 538, pp. 171–183
Kearns, J., Scully, J., Roberge, P., Reichert, D., & Dawson, J. (1996). Electrochemical Noise measurement for Corrosion Applications ASTM.
Lacombe, P., Baroux, B., & Beranger, G. (1993) Stainless steels. Les editions de physique, France
Loto, C., & Cottis, R. (1987). *Corrosion* Vol. 43, pp. 499.
Majidi, A., & Streicher, M. (1984). *Corrosion* Vol. 40, pp. 584
McGuire, G., Azar, N., & Shelhamer, M. (1997). Recurrence matrices and the preservation of dynamical properties. *Phys Lett A* Vol. 237:43–7. 29 December.
McMurray, H., Magill, S., & Jeffs, B. (1996). "Scanning Reference Electrode Technique as a tool for investigating localised corrosion phenomena in galvanised steels", *Iron and Steelmaking*, Vol. 23, No. Issue 2, pp. 183
McMurray, H., & Worsley, D. (1997). "Scanning Electrochemical Techniques for the Study of Localised Metallic Corrosion", in *Research in Chemical Kinetics*, Vol. 4, pp. 149, Compton & Hancock eds. Blackwell Science Ltd.
Marshall, P. (1984). "AUSTENITIC STAINLESS STEELS, Microstructure and Mechanical Properties", Elsevier Applied Science Publishers LTD, England.
Marwan, N., Carmen, M., Thiel, M., & Kurths, J. (2007). Recurrence plots for the analysis of complex systems, *Physics Reports* Vol. 438 pp. 237–329
Monkhorst, H. & Pack, J. (1976). *Phys. Rev. B.* Vol. 13, pp. 5188-5192.
Nakayama, S., Kaji, T., Shibata, M., Notoya, T., & Osakai, T. (2007). *J Electrochem Soc Vol.* 154, C1
Nakayama, S., Kimura, A., Shibata, M., Kiwabata, S., & Osakai T. (2001). *J Electrochem Soc* Vol. 148, B467
Newman, R. (2001). W.R. Whitney award lecture: understanding the corrosion of stainless steel, *Corrosion* Vol. 57, pp. 1030–1041.
Novak, P., Štefec, R., & Franz, F. (1975). *Corrosion* Vol. 31, pp. 344
Ordejon, P., Artacho, E., & Soler, J. (1996). *Phys. Rev. B.* Vol. 53, pp. R10441-R10444.
Perdew, J., Burke, K., & Ernzerhof, M. (1996). *Phys. Rev. Lett.* Vol. 77, pp. 3865-3868.

Peters, K., Walker, C., Steadman, P., Robach, O., Isern, H., & Ferrer, S. (2001). The adsorption of carbon monoxide on Ni(110) above atmospheric pressure investigated with surface X-ray diffraction. *Phys. Rev. Lett.* Vol. 86, pp. 5325

Rodriguez, N., Kim, M., Fortin, F., Mochida, I., & Baker, R. (1997). Carbon deposition on iron-nickel alloy particles, *Appl. Catal., A* Vol. 148, pp. 265–282.

Rosenfeld, I., & Danilov, I. (1967). "Electrochemical Aspects of Pitting Corrosion" *Corrosion Science*, Vol.7, pp.129.

Sánchez-Portal, D., Ordejon, P. & Canadell, E. (2004). Structure and Bonding, Vol. 113, pp. 103-170.

Sankey, O., & Niklewski, D. (1989). *Phys. Rev. B.* Vol. 40, pp. 3979-3995.

Sargeant, D., Hainse, J., & Bates, S. (1989). " Microcomputer controlled scanning reference electrode apparatus developed to study pitting corrosion of gas turbine disc materials ", *Materials Science and Technology*, Vol. 5, pp. 487

Sazou, D., & Pagitsas, M. (2003). Non-linear dynamics of the passivity breakdown of iron in acidic solutions. *Chaos Soliton Fract* Vol. 17:505–22.

Searson, P., & Dawson, J. (1988). *J. Electrochem. Soc.*, Vol. 135, No. issue 8, pp. 1908

Sudesh, T., Wijesinghe, L., & Blackwood D. (1989). Real time pit initiation studies on stainless steels: The effect of sulphide inclusions, *Corros. Sci.* Vol. 49 (2007) 1755 - 1764.

Soler, J., Artacho, E., Gale, J., Garcia, A., Junquera, J., Ordejon, P., & Sanchez, D. (2002). *J. Phys.: Condens. Matter* Vol. 14, pp. 2745-2779

Terada, M., Saiki, M., Costa, I., & Fernando, A. (2006). *J Nucl Mater* Vol. 358, pp. 40

Torres, V., Rodríguez, F., García-Ochoa, E. & Genesca, J. (2006). *Anticorros Meth Mater* Vol. 53, pp. 348

Trethewey, K., Sargeant, D., Marsh, D., & Haines, S. (1994). "New methods on quantitativeanalysis of localised corrosion using scanning electrochemical probes", in Modelling Aqueous Corrosion: From individual Pits to System management, p. 417, eds. K.R.

Trethewey, K., Sargeant, D., Marsh, D., & Tamimi, A. (1993). "Application of the Scanning Reference Electrode Technique to localised corrosion", *Corrosion Science*, Vol.35, No. issue 1-4, pp. 127.

Trethewey, K., Marsh, D., & Sargeant, D. (1996). "Quantitative Measurements of Localised

Corrosion Using SRET", *CORROSION-NACE 94*, Paper no. 317.Troullier, N., & Martins, J. (1991). *Phys. Rev. B.* Vol. 43, pp. 1993-2006.

Trulla, L., Giuliani, A., Zbilut, J., & Webber, C. (1996). *Phys Lett. A* Vol. 223: 255

Tuck, C. (1983). "The Use of Micro-electrodes in the Study of Localised Corrosion in Aluminium Alloys ", *Corrosion Science*, Vol. 23, No. issue 4, pp. 379

Turnbull, A., McCartney, L., & Zhou S. (2006). A model to predict the evolution of pitting corrosion and the pit-to-crack transition incorporating statistically distributed input parameters, *Corros. Sci.* Vol. 48, pp. 2084–2105.

Zaveri, N., Sunb, R., Zufelt, N., Zhoua, A., & Chenb, Y. (2007). *Electrochim Acta* Vol. 52, pp. 5795.

Zbilut, J., & Webber, C. (1992). Embedding and delays as derived from quantification of recurrence plot. *Phys Lett A* Vol. 171, pp. 199–203.

Zhou, S. & Turnbull, A. (1999). Influence of pitting on Fatigue life of Turbine Steel" *Fatigue Fract. Engng. Mater. Struct.* Vol. 22, pp. 1083 -1093.

Watanabe, M., Tomita, M. & Ichin, T. (2001). *J Electrochem Soc* Vol. 148, pp. B522

3

Corrosion Protection of Al Alloys: Organic Coatings and Inhibitors

Ahmed Y. Musa
Dept. of Chemical and Process Engineering, National University of Malaysia, Malaysia

1. Introduction

Aluminium and its alloys have excellent durability and corrosion resistance, but, like most materials, their behaviour can be influenced by the way in which they are used. Aluminium is commercially important metal and its alloys are widely used in many industries such as reaction vessels, pipes, machinery and chemical batteries because of their advantages. Aluminium is a well-known sacrificial anode if couple with more passive metal as it is most reliable and cost effective anode. Aluminium sacrificial anode has been used in major project all over the world. It is used in offshore application including structures, platforms, pipelines, jetties and power plants. Aluminium anode is also used for ship-hull and ballast-tank protection.

Aluminium is an active metal and its resistance to corrosion depends on the formation of the protective oxide film (Sastri et al. 2007). There are several methods to protect the aluminium and its alloys from corrosion such as coatings (Metallic, Inorganic, conversion and organic coatings), control of environment (operating variable i.e. pH, dissolved oxygen, temperature and etc.) and corrosion inhibitors (organic and inorganic additives).

This chapter describes the corrosion and corrosion prevention of the aluminium alloys by organic coatings and inhibitors. Industrial applications and common corrosion form of different Al alloys are carried out in this chapter. The corrosion prevention methods for different Al alloys are also mentioned in this chapter. A recent literature review of more than 30 papers is summarized at the end of this chapter.

2. Industrial applications of different Al alloys

Aluminium can be alloyed with different elements like zinc, magnesium, silicon, copper, manganese, as well as lithium. As a result, it can be used for different applications like manufacturing of aluminium foil covering, food packaging industry, food and chemical industry, vehicle panelling, mine cages, air frames, chemical plants, pressure vessels, road tankers, transportation of ammonium nitrate, irrigation pipes and window frames. Some industrial applications for aluminium alloy are listed below:

1. Aluminium alloys are highly resistant to non-heat treatments. They are good conductors of heat and electricity and that is why they are being used in different chemical industries for preparation of aluminium products.

2. Aluminium alloys get hardened during the process of reactions. That is why they are highly favourable alloys for the factor of weld ability as well as formability. Also, they are superior for the cryogenic uses even in the condition of annealed treatments.
3. Aluminium alloys is its high resistance to corrosion. They deter it from different harsh chemical treatments and help in retaining its lustre and strength. It also has high resistance to sea and ocean water. Therefore, it can be used for different air cages without any hassles.
4. Aluminium alloys can be converted into any form as they are ductile in nature. Be it sheets or wires, they can be drawn into various shapes without any inconvenience.
5. Aluminium alloy also used as sacrificial anode for the cathodic protection system (pipeline cathodic protection, oil tank, ship hull aluminium cp anode, Figure 1, boiler anodes, aluminium rod and aluminium bracelet anodes).

Fig. 1. Sacrificial aluminium anode attached to the hull of a ship.

3. Most common corrosion mechanism for Al alloys

Many different corrosion mechanisms exist for Al alloys. The most common types are generally well understood. For each, the process is complex, incorporates many factors, and varies according to metal and specific operating conditions. Yet all still remain difficult to control, and represent a very serious threat to most industries. Once established, most corrosion problems will produce future years of operating difficulty and expense at varying levels of severity.

3.1 Galvanic corrosion

Galvanic corrosion occurs, when a metallic contact is made between a nobler and a less noble one (Wallen 1986; Dexter 1999; Bardal et al. 1993). A necessary condition is that there is also an electrolytic condition between the metals, so that a closed circuit is established.

The area ratio between cathode and anode is very important. For instance, if the nobler cathodic metal has a large surface area and the less noble metal has a relatively small area, a large cathodic reaction must be balanced by a correspondingly large anodic reaction concentrated in a small area resulting in a higher anodic reaction rate (Wallen 1986). This leads to a higher metal dissolution rate or corrosion rate. Therefore, the ratio of cathodic to anodic area should be kept as low as possible. Galvanic corrosion is one of the major practical corrosion problems of aluminium and aluminium alloys (Dexter 1999) since aluminium is thermodynamically more active than most of the other common structural materials and the passive oxide which protects aluminium may easily be broken down locally when the potential is raised due to contact with a nobler material. This is particularly the case when aluminium and its alloys are exposed in waters containing chlorides or other aggressive species such as SO_4 (Bardal et al. 1993). The series of standard reduction potentials of various metals can be used to explain the risk of galvanic corrosion; however these potentials express thermodynamic properties, which do not take into account the kinetic aspects (Valen et al. 1989). Also, if the potential difference between two metals in a galvanic couple is too large, the more noble metal does not take part in corrosion process with its own ions. Thus, under this condition, the reduction potential of the more noble metal does not play any role. Therefore establishing a galvanic series for specific conditions becomes crucial.

3.2 Pitting

Pitting is a highly localized type of corrosion in the presence of aggressive chloride ions. Pits are initiated at weak sites in the oxide by chloride attack. Pits propagate according to the reactions

$$A1 \rightarrow A1^3 + +3e - \quad (1)$$

$$A1^{3+} + 3H_2O \rightarrow A1(OH)_3 + 3H^+ \quad (2)$$

while hydrogen evolution and oxygen reduction are the important reduction processes at the intermetallic cathodes, as sketched in figure 1:

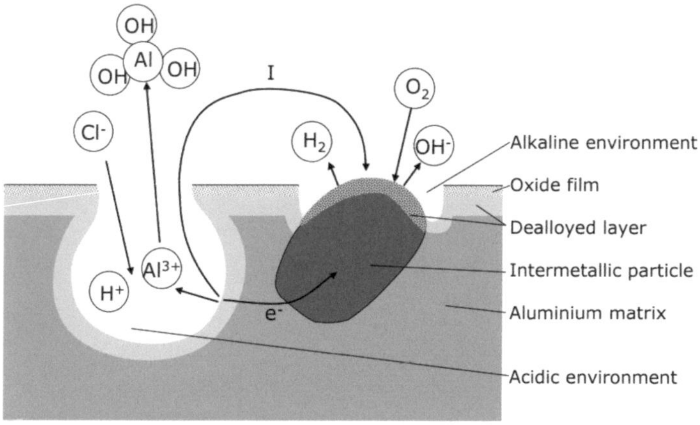

Fig. 2. Pitting corrosion mechanisms for the aluminium.

$$2H^{+} + 2e \rightarrow H_2 \quad (3)$$

$$O_2 + 2H_2O + 4e \rightarrow 4OH \quad (4)$$

As a pit propagates the environment inside the pit (anode) changes.

According to reaction 2 the pH will decrease. To balance the positive charge produced by reaction 1 and 2, chloride ions will migrate into the pit. The resulting HCl formation inside the pit causes accelerated pit propagation. The reduction reaction will cause local alkalinisation around cathodic particles. As previously mentioned aluminium oxide is not stable in such environment, and aluminium around the particles will dissolve (alkaline pits). The active aluminium component of the particles will also dissolve selectively, thereby enriching the particle surface with Fe and increasing its cathodic activity. Etching of the aluminium matrix around the particles may detach the particles from the surface, which may repassivate the alkaline pits. This may also reduce the driving force for the acidic pits causing repassivation of some in the long run. Figure 3 show pitting on an Al alloy.

Fig. 3. SEM images showing the pitting corrosion for the aluminium

3.3 Intergranular corrosion

Intergranular corrosion (IGC) is the selective dissolution of the grain boundary zone, while the bulk grain is not attacked. IGC is also caused by microgalvanic cell action at the grain boundaries. The susceptibility to IGC is known to depend on the alloy composition and thermomechanical processing. Grain boundaries are sites for precipitation and segregation, which makes them physically and chemically different from the matrix. Precipitation of e.g. noble particles at grain boundaries depletes the adjacent zone of these elements, and the depleted zone becomes electrochemically active. The opposite case is also possible; precipitation of active particles at grain boundaries would make the adjacent zone noble. These two cases are illustrated in figure 3. Figure 4 shows examples of intergranular corrosion on 7075-T6 aluminium alloy.

3.4 Exfoliation corrosion

Exfoliation is yet another special form of intergranular corrosion that proceeds laterally from the sites of initiation along planes parallel to the surface, generally at grain boundaries, forming corrosion products that force metal away from the body of the material, giving rise to a layered appearance. Exfoliation is sometimes described as lamellar, layer, or stratified corrosion. In this type of corrosion, attack proceeds along selective subsurface paths parallel to the surface. It is possible to visually recognize this type of corrosion if the grain boundary attack is severe otherwise microstructure examination under a microscope is needed. In Al alloy, exfoliation corrosion occurred when the metal exposed to tropical marine environment. Also note the paint failures caused by corrosion of aluminium at the coating/aluminium interface. Exfoliation corrosion can be prevented by coatings, selecting a more exfoliation resistant aluminium alloy and using heat treatment to control precipitate distribution.

3.5 Stress-Corrosion Cracking (SCC)

Stress-corrosion cracking in aluminium alloys is characteristically intergranular. According to the electrochemical theory, this requires a condition along grain boundaries that makes them anodic to the rest of the microstructure so that corrosion propagates selectively along them. Intergranular (intercrystalline) corrosion is selective attack of grain boundaries or closely adjacent regions without appreciable attack of the grains themselves

Aluminium alloys that contain appreciable amounts of soluble alloying elements, primarily copper, magnesium, silicon, and zinc, are susceptible to stress-corrosion cracking (SCC). An extensive failure analysis shows how many service failures occurred in the industry and what kind of alloys and stresses led to initiation and propagation of stress corrosion cracks which caused these service failures. Alloys 7079-T6, 7075 -T6 and 2024 - T3 contributed to more than 90% of the service failures of all high-strength aluminium alloys. Aluminium and its alloys can fail by cracking along grain boundaries when simultaneously exposed to specific environments and stresses of sufficient magnitude. Well-known specific environments include water vapour, aqueous solutions, organic liquids and liquid metals. Stresses sufficient for crack initiation and crack growth can be far below the stresses required for gross yielding, especially in those alloy/environment combinations that are of practical importance, e.g., high strength aluminium alloys in air. This phenomenon of environment-induced intergranular cracking is often called stress-corrosion cracking. With most service failures specific causes for initiation or propagation of stress corrosion cracks have been observed. The various causes usually belong to one of the following three classes: metallurgical, environmental and mechanical. This follows quite naturally from the old observation that for stress corrosion cracking to occur, three conditions have to be fulfilled: the alloy must be "susceptible" to SCC, the environment must be "damaging" and the stress (intensity) must be "sufficient".

The electrochemical theory of stress corrosion, developed about 1940, describes certain conditions required for SCC of aluminium alloys. Further research showed inadequacies in this theory, and the complex interactions among factors that lead to SCC of aluminium alloys are not yet fully understood. However, there is a general agreement that for aluminium the electrochemical factor predominates and the electrochemical theory continues to be the basis for developing aluminium alloys and tempers resistant to SCC.

4. Corrosion protection of Al alloy

There are several methods commonly used to combat corrosion. These include passive film formation, chromating, cathodic protection, organic coatings and inhibitors (Jones 1996). This chapter is only concerned about the prevention of corrosion using organic coatings and inhibitors

4.1 Organic coatings

Organic coating provides protection either by the formation of a barrier action from the layer or from active corrosion inhibition provides by pigments in the coating. Surface condition of metal converted to more stable state by coating with organic compounds. These coatings delay the generation of electromotive force, causing the corrosion of the substrate (Schweitzer, 2001). Cathodic deposition of organic coatings has gained worldwide acceptance as a coating process for automotive, appliance and general industrial coatings which has been adopted in technology to provide the first prime coat to a variety of products.

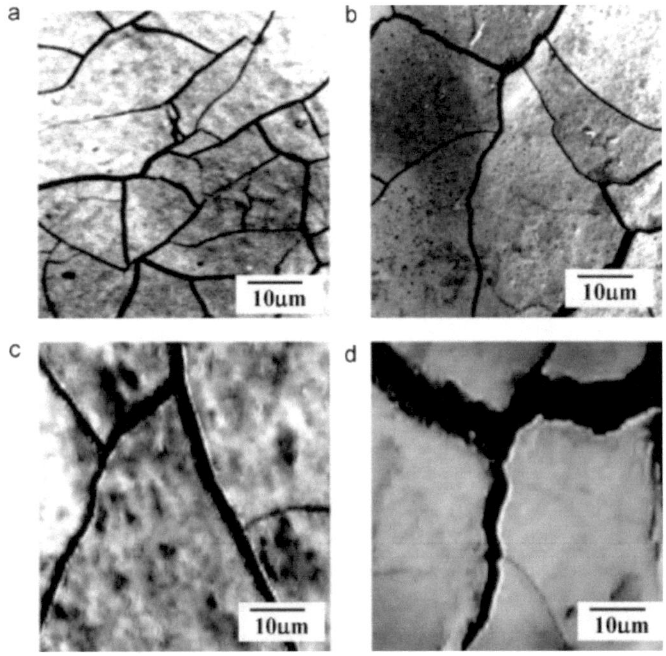

Fig. 4. Optical micrographs of the surfaces of 7075-T6 aluminium alloy specimens exposed to a deaerated 0.5 M NaCl solution at −685 mVSCE: (a) as-received specimen and (b), (c) and (d) specimens with grain sizes of 40, 130 and 290 μm, respectively (El-Amoush, 2011)

Among the large number of electroconducting polymers, polypyrrole and polyaniline are the most promising conducting polymers for corrosion protection. Nevertheless, the lower price of aniline monomer comparing to pyrrole makes polyaniline more challengeable (Popovic and Grgur et. al. 2004). Polyaniline (PANI) exists in a variety of forms that differ in chemical and physical properties. The most common green protonated emeraldine has conductivity on a semiconductor level of the order of 100 S cm^{-1}, many orders of magnitude higher than that of common polymers (<10^{-9} S cm^{-1}) but lower than that of typical metals (>10^{4} S cm^{-1}). Protonated PANI, (e.g., PANI hydrochloride) converts to a nonconducting blue emeraldine base when treated with ammonium hydroxide (Fig. 5). The changes in physicochemical properties of PANI occurring in the response to various external stimuli are used in various applications e.g., in organic electrodes, sensors, and actuators. Other uses are based on the combination of electrical properties typical of semiconductors with materials parameters characteristic of polymers, like the development of "plastic" microelectronics, electrochromic devices, tailor-made composite systems, and "smart" fabrics. The establishment of the physical properties of PANI reflecting the conditions of preparation is thus of fundamental importance (Stejskal & Gilbert 2002). Recently many attempts have been carried out to protect the aluminium and its alloy by organic base coating. Ogurtsov et al. (2004) have reported the protection ability of pure undoped PANI (emeraldine base) and PANI doped with p-toluene-sulfonic (TSA), camphorsulfonic (CSA) and dodecylbenzenesulfonic (DBSA) acids coatings for Al 3003 alloy. As it can be seen from their results, Table 1, that the highest protecting ability factor was obtained for undoped PANI being equal to 12 and 4.4 in neutral and acidic media, respectively.

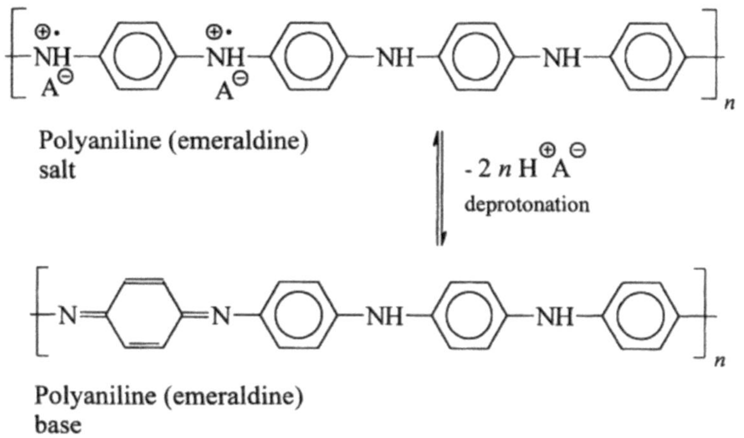

Fig. 5. Polyaniline (emeraldine) salt is deprotonated in the alkaline medium to polyaniline (emeraldine) base. (A- is an arbitrary anion, e.g., chloride.)

Medium	Coating	$-E_0$ (V)	i_{corr} (μA/cm^2)	γ	S'/S
3.5% NaCl water solution	Control without PANI	0.99	3.6	–	2
	Undoped PANI	0.75	0.3	12	2.5
	PANI-TSA	0.76	0.72	5.0	12.4
	PANI-CSA	0.85	4.55	0.79	1
	PANI-DBSA	0.68	4.2	0.86	1.6
0.1 N HCl water solution	Control without PANI	0.7	3.2	–	2
	Undoped PANI	0.62	0.72	4.4	6.4
	PANI-TSA	0.72	2.3	1.4	21.7
	PANI-CSA	0.58	4.86	0.66	1
	PANI-DBSA	0.68	3.5	0.91	2

Table 1. Results of the electrochemical monitoring of coats doped and undoped PANI on Al 3003 in 3.5% NaCl and 0.1 N HCl solutions

Bajat et al (2008) have studied the electrochemical and transport properties and adhesion of epoxy coatings electrodeposited on aluminium pretreated by vinyltriethoxysilane (VTES) during exposure to 3% NaCl. The electrochemical results showed that the pretreatment based on VTES film deposited from 5% solution provides enhanced barrier properties and excellent corrosion protection. Niknahad et al. (2010) have studied the influence of various blends of hexafluorozirconic-acid (Zr), polyacrylic-acid (PAA) and polyacrylamide (PAM) pretreatment on the performance of an epoxy coated aluminium substrate. They have employed the salt spray, humidity chambers and EIS to characterize corrosion performance of coated substrates with different initial surface pretreatments They have found that among the Zr-based formulations, PAA/Zr and PAA/PAM/Zr showed the best adhesion strength, while the later revealed a good corrosion performance as well. Pirhady et al. (2009) have developed silica-based organic–inorganic hybrid nanocomposite films by sol–gel method for corrosion protection of AA2024 alloy. The sol–gel films were synthesized from 3-glycidoxypropyltrimethoxysilane (GPTMS) and tetraethylorthosilicate (TEOS) precursors. They have utilized the potentiodynamic scanning and salt spray tests to study the corrosion protection properties of the films. Their results indicate that the hybrid films provided exceptional barrier and corrosion protection in comparison with untreated aluminium alloy substrate. Kraljic et al. (2003) have electrochemically synthesised the Polyaniline (PANI) from 0.5M aniline solution in 1.5M H_2SO_4 and 0.5M aniline in 3.0M H_3PO_4 by cycling the stainless steel at potential between 400 and 1000 mV for stainless steel. They tested the coated sample in sulphuric acid and phosphoric acid solution and that PANI layer form in H_3PO_4 acid has better protection than PANI layer form in H_2SO_4. This was due to better protection of steel with $PANIPO_4$ layer which have a higher quality of the oxide film and a smaller area of the PANI-free electrode surface. Ozyilmaz et al. (2004) have coated the thin layer of PANI on the stainless steel in oxalic acid solution containing aniline using potentiodynamic cycle between −0.7 and 1.65V. They have observed that the passivations of the substrate and monomer oxidation prior to film growth for both potential ranges were necessary for the polymerization process. They have investigated the corrosion performance of coated stainless steel using electrochemical impedance spectroscopy in hydrochloric acid. Nyquist curves were revealed that coated stainless steel had different corrosion behaviors according to the potential range used for the polymerization. They have observed that the permeability of PANI film was affected by synthesis condition. Bereket et al. (2005) have successfully synthesized the polyaniline (PANI), poly(2-iodoaniline) (PIANi), and poly(aniline-co-2-iodoaniline) (co-PIANi) using cyclic voltammetry in acetonitrile solution containing tetrabuthylammonium perchlorate (TBAP) and perchloric acid ($HClO_4$) on 304-stainless steel electrodes. Their results showed that the polymer and copolymer were different from that of PANi. PIANi and co-PIANi behaved in a similar manner with regard to the corrosion protection of 304- SS in 0.5 M HCl. They found that this is related with the prevention of cathodic reaction taking place at film–solution interface. PANI coatings were able to provide an effective anodic protection addition to barrier properties for the cathodic reaction. Based on research by Shabani-Nooshabadi et al. (2009), homogeneous and adherent polyaniline coatings can be electrosynthesized on aluminium (Al) alloy 3004 from an aqueous solution containing aniline and oxalic acid using galvanostatic conditions 0.5, 1.5, 5 and $15 mAcm^{-2}$ for 1800 second. They found that the electrochemical polymerization of aniline on Al takes place after the passivation of its surface via the formation of Al oxalate complex and results in the generation of green, uniform and strongly adherent polyaniline coatings. The potentiodynamic polarization and EIS studies reveal that the polyaniline acts as a protective layer on Al against corrosion in 3.5% NaCl solution.

Kenneth et. al. (2003) have electrodeposited the polyaniline films at pure aluminium from a tosylic acid solution. They found that these polymer films exhibited similar characteristics as pure polyaniline electrosynthesized at an inert platinum electrode, when removed from their respective substrates. A galvanic interaction between the polymer and the aluminium was observed giving rise to oxidation of the aluminium substrate and reduction of the polymer. Although the aluminium substrate was oxidized there was little evidence of any corrosion protection by the polyaniline coatings. The coated electrode was tested by polarizing the electrode, they found that attack at the aluminium substrate occurred underneath the polymer, and this indicate that chloride anions diffuse across the polymer to react at the underlying substrate.

Martins et al. (2010) have synthesised the polyaniline films on aluminium alloy 6061-T6 in sulphuric acid by electrodeposition using cyclic voltammetry and potentiostatic polarization. They tested the coated electrode using electrochemical techniques to assess the anticorrosive properties of the coatings. From anodic polarization curves, they found out that the corrosion resistance of the coated alloy was not higher than the bare alloy. The values of the corrosion and pitting potentials remain unchanged and the system undergoing pitting corrosion. Due to the conductive character of the polymer, the tests were not conclusive.

Kamaraj et al. (2009) have electropolymerised aniline on AA 7075 T6 alloy from oxalic acid bath by galvanostatic polarization. FTIR spectroscopy studies revealed the presence of both benzenoid and quinoid structures confirming the presence of partially oxidized polyaniline (Emeraldine salt) which is known for its conducting nature. They evaluate these coatings in 1% NaCl by electrochemical impedance spectroscopy and potentiodynamic polarization. They found that there is poor corrosion resistant behaviour due to galvanic action of polyaniline. However, cerium post-treatment for the polyaniline coatings on AA 7075 T6 alloys exhibited very high corrosion protection performance. This is due to decreased rate of oxygen reduction reaction by forming cerium oxide coating on the pinholes in corrosive media containing chloride anions.

4.2 Corrosion inhibitors

Corrosion inhibitors are organic or inorganic chemical substances which, when added at low concentrations, are capable of preventing or controlling corrosion in 80–90% in most cases (Gonza et. al. 2008). According to ISO 8044, corrosion inhibitor is defined as chemical substance that can reduce the corrosion rate with the present of in small concentration without changing the concentration of corroding agent in a corrosion system. Corrosion inhibitor reduces the rate of corrosion with increase the polarization behaviour of anode and cathode. Apart from that, lower the movement of ion to absorb to metal surface and finally increase the electrical resistance on metal surface (Roberge, 2000).

The shortage of inorganic corrosion inhibitors has increased the demands and interest in organic compounds. Many organic compounds have been reported as effective corrosion inhibitors in the literature, but not a great many are ever used in practical systems. Any practical inhibitor must possess properties beyond its ability to control corrosion including cost, toxicity, and compatibility with the expected environment. Available references to corrosion phenomena in the technical literature appeared by the end of the 18 century. The first patent in corrosion inhibition was given to Baldwin, British Patent 2327. Robinson and Sutherland in 1900, US Patent 640491, were using starch as corrosion inhibitor (Hackerman 1990). Since 1950s, the scientific tone on studying corrosion inhibition mechanism was increased and the conversion of the field from art to science has started with the recognition of adsorption phenomena (Hackerman 1990). Good inhibition in any system requires an understanding of the chemistry of the system with the inhibitor plus the knowledge of the parameters involved. Corrosion inhibition is reversible, and a minimum concentration of the inhibiting compound must be present to maintain the inhibiting surface film. Good circulation and the absence of any stagnant areas are necessary to maintain inhibitor concentration. Specially designed mixtures are required when two or more alloys are presented in a system (Jones 1996).

Organic compounds, mainly containing oxygen, nitrogen and sulphur atoms and having multiple bonds, are recognized as effective inhibitors of the corrosion of many metals and alloys. In different media, for a given metal, the efficiency of the inhibitor depends on the stability of the formed complex and the inhibitor molecule should have centres, which are capable of forming bonds with the metal surface via an electron transfer. Generally, a strong co-ordination bond causes higher inhibition efficiency, the inhibition increases in the sequence $O<N<S<P$ (Musa et al. 2009).

Acid solutions are used for pickling, chemical and electrochemical etching of Al foil and lithographic panels which substitute metallic zinc. Since the metal dissolution in such solutions is rather large, it is necessary to inhibit it by the addition of inhibitors, which should provide a good quality pickled metal surface. For cleaning and descaling; acid solution is also widely used in industry (Yao et. al 2007). Acid cleaning baths are employed to remove undesirable scale from the surface of the metals. Once the scale is removed, the acid is then free for further attack on the metal surface. The use of inhibitor is one of the most practical methods for protection against corrosion, especially in acidic media (Touir et al. 2008).

Umoren et al (2008) have studied the inhibition of aluminium corrosion in hydrochloric acid solution by exudate gum from Raphia hookeri. In this research, the corrosion inhibition of aluminium in HCl solution in the presence of exudate gum from Raphia hookeri at temperature range of 30 to 60^{0}C was studied using weight loss and thermometric techniques. The exudate gum acts as an inhibitor in the acid environment. Results of this research are the inhibition efficiency increases with increase in inhibitor concentration. However, the inhibition efficiency decreases with an increase in temperature. Elewady et al. (2008) have studied the effect of anions on the corrosion inhibition of aluminium in HCl using ethyl trimethyl ammonium bromide as cationic inhibitor. The corrosion inhibition of aluminium in hydrochloric acid solution in the presence of ethyl trimethyl ammonium bromide (ETMAB) at temperature range of 30-50^{0}C was studied using the weight loss and polarization techniques. The effect of addition of anions (I^-, SCN^-, Br^-) is also reported. The inhibition efficiency (%IE) increased with increase in concentration of ETMAB. The addition of the anions increased the inhibition efficiency to a considerable extent. The inhibitive action of ETMAB was discussed on the bases of its adsorption on the metal surface, which follow Freundlish adsorption isotherm. The synergistic effect of ETMAB and anions was discussed. Galvanostatic polarization curves indicated that the inhibitor behaves as mixed-type inhibitor. Pinto et al. (2011) have characterized the inhibitory effect of 4-(N,N-imethylamino) benzaldehyde thiosemicarbazone (DMABT)on the corrosion behaviour of 6061 Al–15 vol. pct. SiC(p) composite and its base alloy at different temperatures in acid mixture medium containing varying concentrations of hydrochloric acid and sulphuric acid using Tafel extrapolation technique and ac impedance spectroscopy (EIS). The have found that inhibition efficiencies increase with the increase in inhibitor concentration, but decrease with the increase in temperature and with the increase in concentration of the acid media. They have determined the thermodynamic parameters for dissolution process and they found that the adsorption of DMABT on both the composite and base alloy was physisorption obeying Freundlich adsorption isotherm. Figure 6 shows the SEM images of the composite in uninhibited and inhibited solutions.

Fig. 6. SEM image of the surface of composite after immersion for 2h in 0.5 M H_2SO_4 at 30°C in (A) absence and (B) presence of 400 ppm of DMABT. (Pinto et al. 2011)

Helal and Badawy (2011) have investigated the corrosion inhibition of Mg–Al–Zn alloy in stagnant naturally aerated chloride freebneutral solutions using amino acids as environmentally safe corrosion inhibitors. They have calculated the corrosion rate in the absence and presence of the corrosion inhibitor using the polarization technique and electrochemical impedance spectroscopy. They have found that Phenyl alanine has corrosion inhibition efficiency up to 93% at a concentration of 2×10^{-3} mol dm^{-3}. The free energy of the adsorption process revealed a physical adsorption of the inhibitor molecules on the alloy surface. Harvey et al. (2011) have studied a range of structurally-related compounds for their capacity to inhibit corrosion on aluminium alloys AA2024-T3 and AA7075-T6 in 0.1 M NaCl solution. Their selected compounds are shown in Figure 7. They found that the thiol group, positions para- and ortho- to a carboxylate, and substitution of N for C in certain positions strongly inhibited corrosion. The hydroxyl group is slightly inhibitive, while the carboxylate group provided little or no corrosion inhibition on its own.

In several cases, different activities are found on the different alloys, with some compounds (particularly thiol-containing compounds) being more effective on AA2024 than on AA7075. Halambek at al. (2010) have reported Lavandula angustifolia L. as corrosion inhibitor of Al-3 Mg alloy in 3% NaCl solution using weight loss, polarization measurements and SEM. The oil dissolved in ethanol and used as 30% (v/v) solution, is found to retard the corrosion rate of Al-3 Mg alloy even at high temperatures. The inhibiting effect of lavender oil components on Al-3 Mg surface is attributed to the Langmuir's adsorption isotherm. It was found that the L. angustifolia L. oil provides a good protection to Al-3 Mg alloy against pitting corrosion in sodium chloride solution, see Figure 7. Noor (2009) has studied the inhibitive action of some quaternary N-heterocyclic compounds namely 1-methyl-4[4(-X)-styryl] pyridinium iodides (X: -H, -Cl and -OH) on the corrosion of Al–Cu alloy in 0.5 M HCl solutions using potentiodynamic polarization, electrochemical impedance spectroscopy and weight loss measurements. She found that the studied compounds are cathodic inhibitors without changing the mechanism of hydrogen evolution reaction. The adsorption of all inhibitors on Al–Cu alloy obeys Langmiur adsorption isotherm. Her data revealed that the studied compounds have good pickling inhibitor's quality as they perform well even at relatively high temperature.

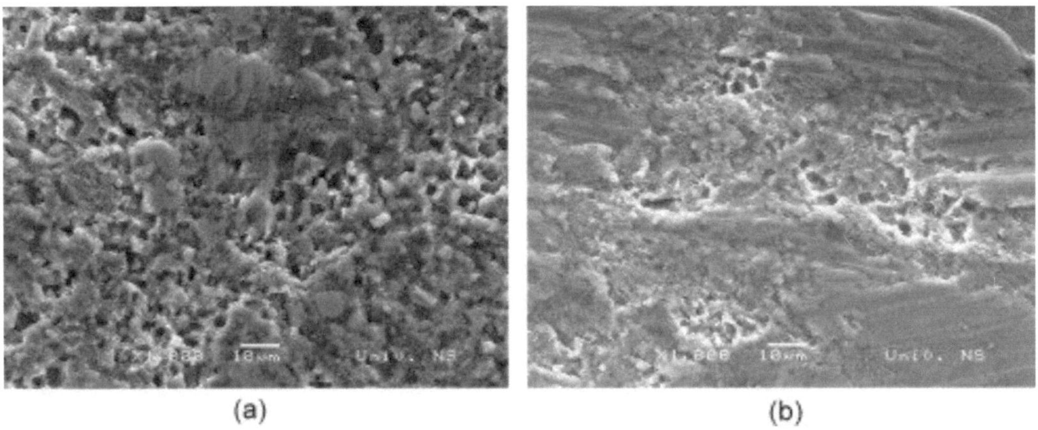

Fig. 7. SEM micrographs of the surface of Al-3 Mg alloy samples after electrochemical tests (polarization measurements) in 3% NaCl solution: (a) without lavender oil; (b) containing 20 ppm lavender oil.

Mishra and Balasubramaniam (2007) have investigated the effect of $LaCl_3$ and $CeCl_3$ inhibitor additions in 3.5% NaCl solution on the corrosion behaviour of aluminium alloy AA2014. Their results showed that polarization resistance increased significantly and the corrosion rate decreased by an order of magnitude with the addition of 1000 ppm of $LaCl_3$ and $CeCl_3$, with maximum decrease noticed for $CeCl_3$. Their SEM monographs confirmed formation of precipitates of oxide/hydroxide of lanthanum and cerium on cathodic intermetallic sites, which reduced the overall corrosion rate, see Figure 8.

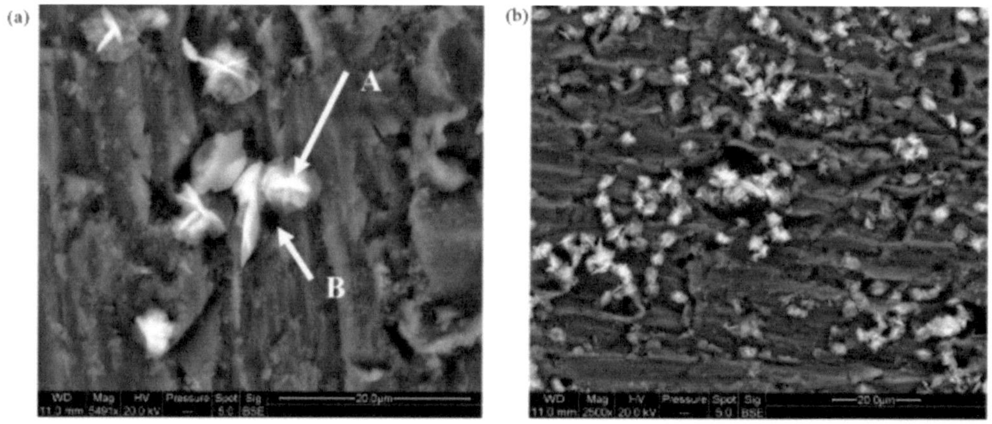

Fig. 8. SEM micrograph of surface after immersion for 4 h in 3.5% NaCl solution (a) with 500 ppm LaCl3, (b) with 1000 ppm $LaCl_3$.

control	Chromate (Cr(VI))	100 ± 0	100 ± 0
1A	4,5-diamino-2,6-dimercaptopyrimidine	87 ± 0	80 ± 0
1B	4,5-diaminopyrimidine	47 ± 3	84 ± 2
1C	Na-(diethyl(dithiocarbamate))	97 ± 1	96 ± 4
1D	2-mercaptopyrimidine	89 ± 4	50 ± 2
1E	Pyrimidine	−153 ± 1	−220 ± 16
1G	Na-benzoate	−80 ± 14	−62 ± 9
1H	Thiophenol	93 ± 5	87 ± 4
1I	Pyridine	−139 ± 18	−150 ± 16
1J	Na-(4-phenylbenzoate)	−72 ± 6	−143 ± 1
1K	Na-4-hydroxybenzoate	−34 ± 5	−56 ± 28
1L	Na-4-mercaptobenzoate	97 ± 2	76 ± 1
1M	Na-(6-mercaptonicotinate)	94 ± 0	86 ± 0
1N	Na-nicotinate	−107 ± 11	-91 ± 26
1O	Na-isonicotinate	−12 ± 1	−45 ± 17
1P	Na-picolinate	58 ± 0	14 ± 2
1Q	Na-3-mercaptobenzoate	16 ± 8	−22 ± 2
1R	Na-salicylate	−175 ± 32	−89 ± 13
1S	Na-2-mercaptobenzoate	88 ± 0	80 ± 0
1T	Na-2-mercaptonicotinate	83 ± 2	70 ± 4
1U	Na-(2,3-mercaptosuccinate)	82 ± 1	48 ± 2
1V	Na-mercaptoacetate	96 ± 1	83 ± 0
1W	Na-mercaptopropionate	100 ± 0	31 ± 9
1X	Na-acetate	−12 ± 8	15 ± 8

Table 2. Compounds in Fig. 8 tested for corrosion inhibition of AA2024 and AA7075 in 0.1 M NaCl at 21 °C for 4 weeks. The prefix Na- denotes the mono-sodium salt. Positive inhibitor efficiencies indicate corrosion was retarded while negative values indicate it was accelerated. All compounds were tested at 1 mM, except 1A (0.5 mM) and 2D (0.1 mM) due to their low solubility. (Harvey et al. 2011)

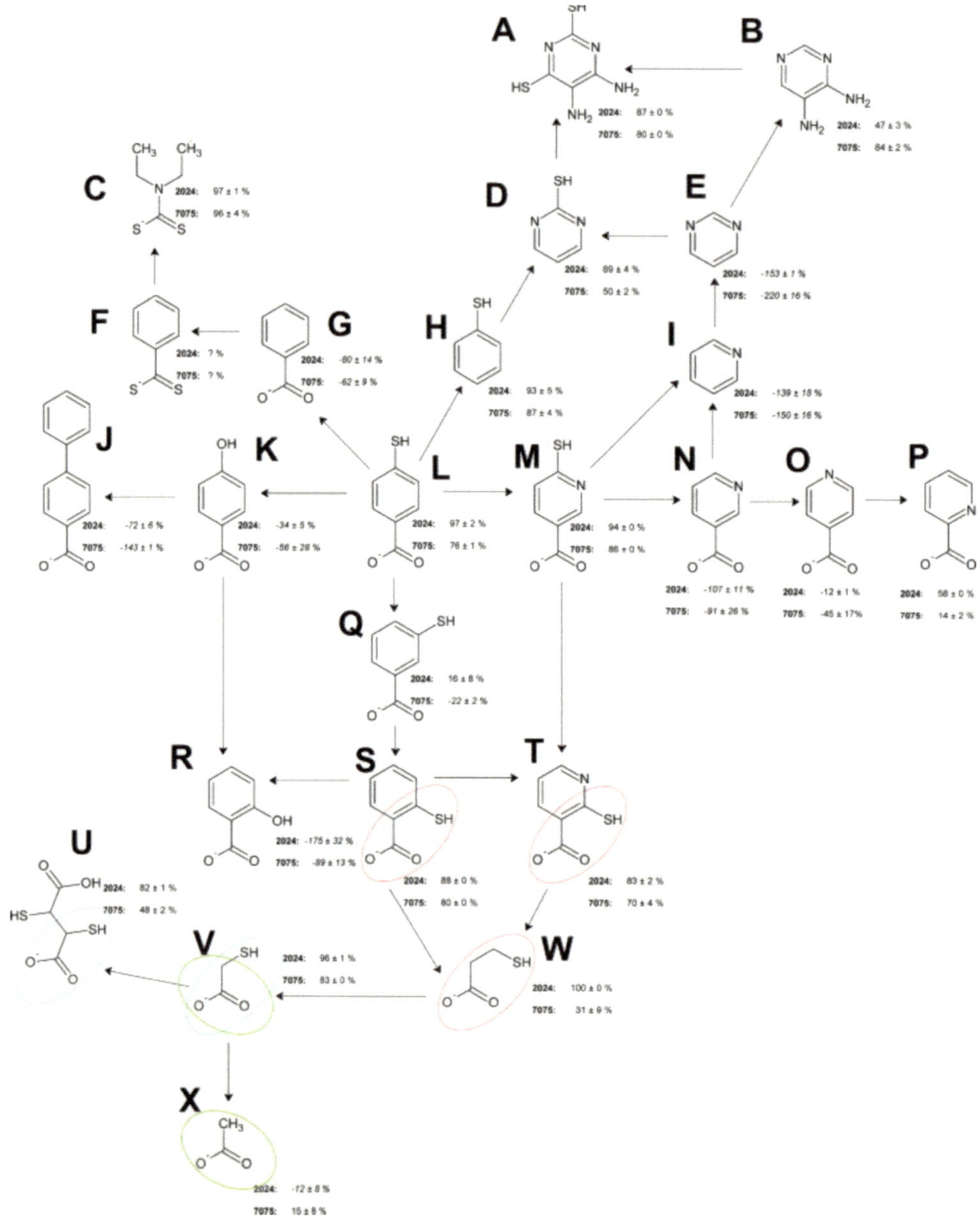

Fig. 9. Illustration of the structures of inhibitors listed in Table 2, showing structural relationships. (Note: similar sub-structures are highlighted with coloured circles.) (For interpretation of the references to colour in this figure legend, the reader is referred to the web version of this paper, (Harvey et al. 2011)

5. Acknowledgment

Author is gratefully acknowledging Universiti Kebangsaan Malaysia for the support of this project under Grant No.UKM-GGPM-NBT-037-2011.

6. References

Bajat, J.B.; Mišković´-Stanković, V.B.; Kacˇarević´-Popović´, Z. (2008). Corrosion stability of epoxy coatings on aluminum pretreated by vinyltriethoxysilane. *Corrosion Science*. Vol. 50 (April 2008), pp. 2078–2084, ISSN 0010-938X.

Bardal, E., Drugli, J. M. & Gartland, P. O.(1993). The behaviour of corrosion resistant steels in seawater: A review. *Corrosion Science*, Vol.30 (April 1993), pp.343-353. ISSN. 0010-938X.

Bereket, G.; Hur, E. & Sahin, Y. (2005). Electrodeposition of Polyaniline, Poly(2-iodoaniline), and Poly(aniline-co-2-iodoaniline) on Steel Surfaces and Corrosion Protection of Steel. *Applied Surface Science*. Vol. 252, (December 2005), pp. 1233–1244, ISSN: 0169-4332.

Dexter, S. C. (1999). Galvanic Corrosion. USA: University of Delaware Sea Grant College Program.

El-Amoush, A. S. (2011). Intergranular corrosion behavior of the 7075-T6 aluminum alloy under different annealing conditions. *Corrosion Science*. Vol. 126 No. 3. (April 2011), pp. 607-613. ISSN 0010-938X.

Elewady, G.Y.; El-Said, I.A. & Foud, A.S.(2008). Effect of Anions on the Corrosion Inhibition of Aluminum in HCl using Ethyl Trimethyl Ammonium Bromide as Cationic Inhibitor. *Int. J. Electrochem. Sci*. Vol. 3 (March, 2008), pp. 644 - 655. ISSN. 1452-3981.

Gonza, C. A.; Francisco J. R; Genesca´ -Llongueras, J. (2008). The influence of Desulfovibrio vulgaris on the Efficiency of Imidazoline as A Corrosion Inhibitor on Low-carbon Steel in Seawater. *Electrochimica Acta*. Vol 54, (December 2008), pp.86-90. ISSN. 0013-4686

Hackerman, N. (1990). Review of corrosion inhibition science and technology - historical perspectives. *Materials Performance*, Vol. 29 (May 1990) pp.44-47. ISSN 0094-1492.

Halambek, J. Berkovic, K. & Vorkapic-Furac, J. (2010). The influence of Lavandula angustifolia L. oil on corrosion of Al-3Mg alloy. *Corrosion Science*. Vol. 52 (December 2010) pp. 3978–3983, ISSN. 0010-938X.

Harvey, T. G.; Hardin, S.G. ; Hughes, A.E.; Muster, T.H. ; White, P.A.; Markley, T.A.; Corrigan, P.A.; Mardel, J.; Garcia, S.J.; Mol, J.M.C. & Glenn, A.M. (2011). The effect of inhibitor structure on the corrosion of AA2024 and AA7075. *Corrosion science*. Vol. 53 (June 2011), pp.2184–2190, ISSN. 0010-938X.

Helal, N. H. & Badawy, W.A. (2011). Environmentally safe corrosion inhibition of Mg–Al–Zn alloy in chloride free neutral solutions by amino acids. *Electrochimica Acta*. Vol. 56 (July 2011), pp. 6581- 6587, ISSN. 0013-4686

Jones, D. A. (1996). *Principles and Prevention of Corrosion*. ISBN. 0133599930. Prentice-Hall International. London, UK.

Kamaraj. K; Sathiyanarayanan, S. & Venkatachari, G. (2009). Electropolymerised Polyaniline Films on AA 7075 Alloy and its Corrosion Protection Performance. *Progress in Organic Coatings*. Vol. 64 ( ), pp. 67–73, ISSN 0300-9440

Kenneth, G; Conroy, C. & Breslin, B. (2003). The Electrochemical Deposition of Polyaniline at Pure Aluminium: Electrochemical Activity and Corrosion Protection Properties. *Electrochimica Acta*. Vol. 48, (February 2003), pp. 721-732, ISSN.0013-4686

Kraljic. M, Z. Mandic, Lj. Duic. (2003). Inhibition of Steel Corrosion by Polyaniline Coatings. *Corrosion Science*. Vol. 45 (January 2003), pp. 181–198, ISSN 0010-938X.

Mishra, A. K. & Balasubramaniam, R. (2007). Corrosion inhibition of aluminum alloy AAm2014 by rare earth chlorides. *Corrosion Science*. Vol. 49 (June 2007) pp. 1027-1044, ISSN. 0010-938X.

Musa A. Y.; Kadhum, A. A. H.; Daud, A. R.; Mohamad, A. B.; Takriff, M. S. & Kamarudin, S. K. (2009). A comparative study of the corrosion inhibition of mild steel in sulphuric acid by 4, 4-dimethyloxazolidine-2-thione. *Corrosion Science*. Vol. 51, (October 2009), pp. 2393–2399, ISSN. 0010-938X.

Niknahad, M. , Moradian, S.& Mirabedini, S. M. (2010). The adhesion properties and corrosion performance of differently pretreated epoxy coatings on an aluminium alloy. *Corrosion Science*. Vol. 52 (February 2010), pp. 1948–1957, ISSN 0010-938X.

Noor, E. A. (2009). Evaluation of inhibitive action of some quaternary N-heterocyclic compounds on the corrosion of Al–Cu alloy in hydrochloric acid. *Materials Chemistry and Physics*. Vol. 114 (April 2009) pp. 533–541, ISSN. 0254-0584

Ogurtsov, N. A.; Pud, A. A.; Kamarchik, P & Shapoval, G. S. (2004). Corrosion inhibition of aluminum alloy in chloride mediums by undoped and doped forms of polyaniline. *Synthetic Metals*. Vol. 143, No.1, (May 2004), pp. 43-47, ISSN. 0379-6779.

Ozyilmaz A.T; Erbil, M.; Yazici. B. (2004) . Investigation of Corrosion Behaviour of Stainless Steel Coated with Polyaniline via Electrochemical Impedance Spectroscopy. *Progress in Organic Coatings*. Vol. 51 (October 2004), pp. 47–54, ISSN 0300-9440.

Pierre, R. R. (2000). *Handbook of Corrosion Engineering*. McGraw-Hill, ISBN. 0070765162, New Jersy, USA

Pinto, G. M.; Nayak, J. & Shetty, A. N. (2011). Corrosion inhibition of 6061 Al–15 vol. pct. SiC(p) composite and its base alloy in a mixture of sulphuric acid and hydrochloric acid by 4-(N,N-dimethyl amino) benzaldehyde thiosemicarbazone. *Materials Chemistry and Physics*. Vol. 125 (February 2011), pp. 628–640, ISSN. 0254-0584.

Pirhady Tavandashti, N.; Sanjabi, S. & Shahrabi, T. (2009). Corrosion protection evaluation of silica/epoxy hybrid nanocomposite coatings to AA2024. *Progress in Organic Coatings*. Vol. 65 (October 2008), pp. 182–186, ISSN 0300-9440.

Popovi, M. M. & Grgur, B. N.(2004). Electrochemical synthesis and corrosion behavior of thin polyaniline-benzoate film on mild steel. *Synthetic Metals*. Vol. 143, No. 2, (June 2004), pp. 191-195, ISSN 0379-6779

Sastri, V. S.; Ghali, E.& Elboujdaini, M. (2007). *Corrosion Prevention and Protection Practical Solutions*, John Wiley & Sons, ISBN 978-0-470-02402-7, Chichester, England

Schweitzer, P. A. (2007). *Corrosion of Linings and Coatings: Cathodic and Inhibitor Protection and Corrosion Monitoring*. CRC press, ISBN 084-9-382-475, Florida, USA

Stejskal, J. & Gilbert, R. G. (2002) Polyaniline. Preparation of a Conducting Polymer. *Pure Appl. Chem.*, Vol. 74, No. 5 (April 2002), pp. 857–867. ISSN. 0033-4545.

Shabani-Nooshabadi. M, Ghoreishi, S.M. & Behpour, M. (2009). Electropolymerized Polyaniline Coatings on Aluminum Alloy 3004 and their Corrosion Protection Performance. *Electrochimica Acta*. Vol. 54, (November 2009), pp. 6989–6995, ISSN. 0013-4686

Touir, R. Cenoui, M., El Bakri, M. & Ebn Touhami, M. (2008). Sodium gluconate as corrosion and scale inhibitor of ordinary steel in simulated cooling water. *Corrosion Science*. Vol 50 (June 2008) pp.1530-1537. ISSN. 0010-938X.

Umoren, S.A., Obot, I.B.; Ebenso, E.E. & Obi-Egbedi, N.O. (2009). The Inhibition of aluminium corrosion in hydrochloric acid solution by exudate gum from Raphia hookeri. *Desalination*. Vol. 247, No. 1-3, (October 2009), pp. 561-572. ISSN. 0011-9164

Valen, S., Bardal, E., Rogne, T. & Drugli, J. M.(1989). New Galvanic Series Based Upon Long Duration Testing in Flowing Seawater. *11th Scandinavian Corrosion Congress*, pp. 234-240.

Wallen, B. & Anderson, T. (1986). Galvanic Corrosion of Copper Alloys in Contact with a Highly Alloyed Stainless Steel in Seawater. *10th Scandinavian Corrosion Congress*, pp. 149-154.

Yao, S., Jianga, X., Zhoua, L., Lva, Y. & Hub, X. (2007). Corrosion inhibition of iron in 20% hydrochloric acid by 1,4/1,6-bis(α-octylpyridinium)butane/hexane dibromide. *Materials Chemistry and Physics*, Vol. 104 (August 2007), pp.301-305. ISSN: 0254-0584

4

The Role of Silica Fume Pigments in Corrosion Protection of Steel Surfaces

Nivin M. Ahmed[1] and Hesham Tawfik M. Abdel-Fatah[2]
[1]*Polymers and Pigments Department, National Research Centre, Dokki, Cairo,*
[2]*Central Chemical Laboratories, Egyptian Electricity Holding Company, Sabtia, Cairo, Egypt*

1. Introduction

Inorganic Pigments significantly change our surroundings. They are irreplaceable for serving the purpose of imparting color to various compounds. They also add properties such as rust inhibition, rigidity, and abrasion resistance. Pigments are insoluble substances that can be incorporated into a material to selectively absorb or scatter light. Depending on the specific pigment used, different visual effects are produced. Inorganic pigments may be obtained from a variety of naturally occurring or synthetically produced mineral sources. Their applications range from concrete to artist's colors, from industrial paints to toners in photocopiers, from coloring in foodstuffs to raw materials for catalysts. The application properties of pigments depend not only on their chemistry but also on their chemical appearance and to greater extent on the manufacture process (Barnett et al., 2006; Talbert, 2007; Brock, 2000; Baxbaum & Pfaff, 2005).

In comparison with organic pigments, inorganic pigments are generally better to withstand the effects of sunlight and chemical exposure. They provide superior opacity, which means they can render a substance or object opaque by prohibiting light from passing through it. Inorganic colors, however, tend to be less bright, pure, and rich than their organic counterparts. Inorganic pigments also possess less tinting strength; i.e. more pigment is needed to produce the desired effect. This generally makes them more durable. Almost all inorganic pigments are completely insoluble. Consequently they do not bleed or leach out of coatings, inks, or plastics. In addition, inorganic pigments are usually less expensive than similar organic colors (Gysau, 2006; Brock et al., 2000; Baxbaum & Pfaff, 2005; Degryse, 2003).

Nowadays there is evolution in pigment industry, manufacturers are trying to evolve new trends of pigments which are safe, cheap with high efficiency to replace the common and well known pigments, among these new trends of pigments are;

a. **Substrate pigments**, where at least one additional component (pigment or extender) is deposited onto a substrate (pigment or extender), preferably by a wet method. Weak, medium or strong attractive forces develop between these pigment components during drying or calcining. These forces prevent segregations between components during use (Baxbaum & Pfaff, 2005; Degryse, 2003; Herbst& Hunger 2006).

b. **Special substrate pigments** include the after-treated pigments and the core pigments. To produce the after-treated pigments the inorganic pigment particles are covered with a thin film of inorganic or organic substances to suppress undesirable properties (e.g. catalytic and photochemical reactivity) or to improve the dispersibility of the pigments and the hydrophilic or hydrophobic character of their surfaces. The particles can be coated by precipitation, by adsorption of suitable substances from solutions (usually aqueous), or by steam hydrolysis (Baxbaum & Pfaff, 2005; Degryse, 2003; Herbst& Hunger 2006; Warson & Finch, 2001; Forsgren, 2006; Vesely &Kalendova, 2008). To produce core pigments. A pigment is deposited on an extender, by precipitation or by wet mixing of the components. In the case of anticorrosive pigments, whose protective effect is located on their surfaces, the use of core pigments can bring about a significant saving of expensive materials. Extender particles are also treated by fixing water-insoluble organic dyes on their surfaces via lake formation (Vesely &Kalendova, 2008; Veleva, 1999).

Titanium dioxide and silica are among the most frequently used fillers in paint and coating formulations. The special properties of the materials should improve the product formulation.

Silica fume is a byproduct of producing silicon metal or ferrosilicon alloys, Silicon metal and alloys are produced in electric furnaces. The smoke that results from furnace operation in tonnage amounts is collected and sold as silica fume, rather than being land-filled. Nowadays, there is increasing environmental concern with regard to excessive volumes of solid waste hazards accumulation. Silica fume which is a fine lightweight fluffy amorphous powder that possesses suitable values of specific gravity, bulking value and oil absorption can find a new market in different industries (Ayana et al, 2005). Silica fume is inert, neutral and of an excellent chemical resistance. It consists primarily of amorphous (non-crystalline) silicon dioxide (SiO_2); its amorphous nature favors safe use from a standpoint of industrial hygiene. It has spherical fine particles less than 0.01μ, the average particle size ranges between 0.1-0.3 μ, easy to lump together to form a looser, with high adsorption capacity and gel. Its average granule diameter is 0.15~0.20 μm, and specific surface area is 15000~20000 m^2/kg. Major Features of silica powder are its high purity (SiO_2 Content Above 99.5%), hardness 7, high White Reached 96 Degrees, reasonable oil absorption, good dispersion, and stable chemical properties. It improves coatings corrosion resistance, wear resistance, and high temperature resistant properties (Del Amo et al, 2002; Perrera, 2004; Molera et al., 2004; Sorensen et al., 2009).

Silica fume is widely used in hydropower projects, refractories, roads, bridges, tunnels, chemical ceramic, rubber and other industries. Because of its fine particles, large surface area, and the high SiO_2 content, silica fume is very reactive material. It can be successfully applied in raising intensity, durability, and it also can improve the material overall performance as filler, used in paint, coating, rubber and other high molecular material (kirubaharan et al., 2009; Guin et al., 2011; Siddique & Khan, 2011, Weil, 2011).

The health problems concerning fumed silica are low; it is not listed as a carcinogen by OSHA, IARC, or NTP. Due to its fineness and thinness, fumed silica can easily become airborne, making it an inhalation risk, capable of causing irritation.

The extremely fine and non-porous nature of pyrogenic silica or fumed silica as it is also known, make it an ideal thickening and bulking agent with outstanding thixotropic qualities. Thixotropic refers to a substances characteristic of reducing viscosity or thickness with extended agitation or shaking. This makes fumed silica ideal filler for paints, thus causing them to thin out during application and regain their viscosity when left to stand preventing drips and runs. This characteristic is also beneficial in the formulation of printing inks which allow for high definition levels during application (Yingchao et al., 2011; Otterstedt & Bradreth, 1998).

Silica fume products have very high purity resulting in highly consistent physical properties and particle size distributions. Their complete inertness and neutral pH, make these products not capable of altering or initiating reactions when incorporated in catalyzed or multi-component chemical systems. Silica fume filler can be used in the following coating and paint industry field (Baxbaum & Pfaff, 2005; Forsgren, 2006; Patton, 1973):

- Industry painting
- Wooding coating
- Anti-corruption coating
- Powder coating
- Architectural Paint
- Water -Proof Paint

Silica fume can benefit (Ayana, 2005; Kirubaharan, 2009):

- For improved abrasion resistance and scratch resistance of the film.
- Gives the highest chemical durability in industrial coatings.
- Offers high loading level leads to reduced formulation cost.

The modern coatings industry utilizes evolving chemistries in its quest to become more efficient in protecting and enhancing the various substrates to which coatings are applied. Further, in order to limit toxic and environmentally damaging emissions, the industry is striving to move away from solvents to water-based employing eco-friendly pigments. This is focused on providing some of the chemical solutions to the many challenges the industry face today in generating modern, safe, low-emission paints and coatings.

Titanium dioxide is a chemically very inert and light resistant material. TiO_2 is mainly known as white pigment (so-called titanium white) and is valued for its brightening effects, and its dirt-repellent properties. TiO_2 particles are used in the coatings industry because of its strong color, high opacity due to their high refractive index, improving the light resistance and dirt resistance of a formulation, low oil absorption, high tinting strength, inert chemical properties and recently attractively as photo-catalyst derived from their photo-catalytic activity. Although high photo-catalytic activity is required for titanium dioxide used as photo-catalysts, low photo-catalytic activity is sometimes preferred for titanium dioxide additives to avoid the degradation of matrix. The catalytic activity of titanium dioxide particles can be enhanced or suppressed by precipitating them on a core of other appropriate pigment to control this phenomenon. The insulating layers like silica (SiO_2), alumina (Al_2O_3), or a polymer are often expected to suppress the catalytic activity of titanium dioxide particles (Peter & Robert, 1999; Salamone, 1999; Streitberger & Kreis, 2008; Martens, 1974; Bremmell & Mensah, 2005).

The refractive index is a measure of how light is bent when it passes from one medium to another. The higher the refractive index, the more the light is bent which results in greater opacity. Rutile TiO_2 has a high refractive index and gives good opacity to coatings. Most mineral fillers have a significantly lower refractive index than TiO_2 and don't contribute to the opacity, but they can be used in conjunction with TiO_2 to achieve opacity at reduced cost. Some minerals, such as amorphous silica (silica fume), have refractive index the same or lower than the resin and will be invisible in the dry film. They can be used to reduce the gloss of a clear coating without creating haze (Bremmell & Mensah, 2005; Shao et al., 2009; Gaumet et al., 1997; Walter, 1991).

This work addresses preparation of new pigments via new method of preparation named "Core-shell", where silica fume which is a cheap and waste material representing the core comprises 90% of the new pigments is covered with a thin layer of titanium dioxide that does not exceed 10% of the pigment comprising the shell. This method of preparation presents a new easy route to obtain high performance pigments with concomitant savings besides being eco-friendly. Presence of titanium dioxide is expected to increase the hiding power of silica fume and consequently increases the new pigments opacity. The combination of these two compounds can lead to the production of new pigments with improved properties different from each of its individual components, overcoming their deficiencies; and consequently changing their efficiency of protection when applied in paints. Titanium dioxide was deposited on silica fume surface in three different concentrations to study the effect of its presence concentration on the anticorrosion efficiency of the new core-shell pigments.

2. Experimental

2.1 Materials

All the employed pigments, extenders, resins, solvents, additives and chemicals were products of different local and international companies.

Silica fume waste used has the following chemical composition.

Concentration of main constituents	Wt., %
SiO_2	95.54
Al_2O_3	0.34
Fe_2O_3	0.91
MgO	0.70
CaO	0.32
Na_2O	0.39
K_2O	0.80
P_2O_5	0.07
SO_3	0.03
L.O.I	4.93

Table 1. Chemical composition of silica fume

2.1.1 Preparation of TiO_2/SiO_2 fume core-shell pigments

Titanium tetrachloride was added in three concentrations 1, 2 and 3 ml to 100 ml hydrochloric acid. Silica was immersed in these three solutions and left for sometime to assure complete covering. Ammonia solution was added drop-wisely to these impregnated silicas to adjust their pH; till complete precipitation of the pigments. The formed paste is then filtered through a buchner system and washed very well. This paste is then calcined at 500-750°C. Three concentrations of titanium dioxide and consequently of titanium were deposited on silica surface, these three concentrations are shown in Table (2).

Pigments	Concentration of $TiCl_4$ (wt. %)	Concentration of TiO_2 (wt. %)	Concentration of Ti (wt. %)
TiO_2/SiO_2 (1)	1	0.42	0.25
TiO_2/SiO_2 (2)	2	0.84	0.50
TiO_2/SiO_2 (3)	3	1.26	0.76

Table 2. Concentrations of titanium dioxide and titanium on silica surface

The scheme of preparation was as follows

$$TiCl_4 / SiO_2 \underset{NH_4OH}{\overset{HCl}{\rightarrow}} TiOH_4 / SiO_2 + NH_4Cl \underset{500-750°C}{\overset{\Delta}{\rightarrow}} TiO_2 / SiO_2$$

2.2 Specification of the physical–chemical properties of the prepared pigments

2.2.1 Specific gravity

Determination of specific gravity was carried out according to (ASTM D5965-96, 2007).

2.2.2 Oil absorption

This was carried out by measuring the amount of flax oil in grams that makes a paste with defined properties out of 100 g of pigments. This determination was carried out using the pestle-bowl method according to (ASTM D281-95, 2007).

2.2.3 Bulking value

Determination of bulking value which expresses the volume of paint for a given weight of a pigment, in which pigments with higher bulking values are considered more economic, is carried out according to (ASTM D 16-62, 2007).

2.2.4 Determining pH of the aqueous extracts of the pigments

The procedure of determining aqueous extract pH was derived from a procedure according to (ASTM D 1583-01, 2007). Suspensions of the pigments were prepared in redistilled water.

2.3 Methods of instrumental analysis

2.3.1 X-ray diffraction

X-ray powder diffraction patterns were obtained at room temperature using a Philip's diffractometer (type PW1390), employing Ni-filtered Cu Kα radiation (λ =1.5404 Å). The diffraction angle, 2θ, was scanned at a rate of 2°/ min.

2.3.2 SEM/EDAX analysis

Energy-dispersive X-ray analysis technique, and scanning electron microscopy, (JEOL JX 840), micro-analyzer electron probe, was used in this work to estimate the particle shapes and to determine the elements deposited on silica surface to estimate the formation of the new pigments.

2.3.3 Transmission electron microscopy

Various pigments were examined using (JEOL JX 1230) technique with micro-analyzer electron probe. This technique confirmed the results of the above SEM results besides determining the particle sizes of the prepared pigments.

2.3.4 X-ray fluorescence

The different concentrations of each element in the prepared pigments were determined using Axios, sequential WD-XRF spectrometer, PANalytical 2005.

2.4 Formulation of paints based on alkyd resin

2.4.1 Preparation of anticorrosive paint formulations

The prepared pigments were tested in 15 paint formulations. These paint formulations contain silica fume, commercial titanium dioxide, silica fume covered with different layer thicknesses of titanium dioxide denoted as $SiO_2/TiO_2(1)$, $SiO_2/TiO_2(2)$, and $SiO_2/TiO_2(3)$ expressing concentration of titanium dioxide layers in ascending order from the lower to the higher concentration. All paint formulations were based on medium oil-modified soya-bean dehydrated castor oil alkyd resin. The formulations were divided into three groups; each is based on different P/B ratio (group I has P/B 1.74, group II has P/B 2.175 and group III has P/B 3). Paint formulations are given in Tables 5-7, while the physical-mechanical and corrosion properties are represented in Table 8. The corrosion features of the paint films are shown in Figs. (5-7).

2.5 The effects of the prepared pigments on the mechanical properties of paints

Because the aim of this study was to formulate a pigment whose properties would contribute to the improvement of the mechanical and corrosion qualities of the paints, selected physical–chemical and corrosion tests were carried out.

2.5.1 Determining the resistance of paints against impact (ASTM D 5638-00, 2007)

The result of this test reveals at what height (in cm) of the free fall of a weight onto the paint, the paint film has not yet been disturbed.

2.5.2 Determination of paint resistance against cupping in Erichsen apparatus (ASTM D 5638-00, 2007)

The result of this test reveals the cupping of the test panel with a coating in mm at which the first impairment of the paint occurred.

2.5.3 Determination of paint hardness (ASTM D 6577-00, 2007)

The result of this test indicates the elasticity of paint film by using the simple pendulum test with a needle, when this needle reaches a groove which is made in the paint film, the time it takes in seconds is the measure of the hardness of paint films.

2.5.4 Determining the degree of coating adhesion by means of a cross-cut test (ASTM D 3359-97, 2007)

The test was performed with a special cutting knife whose edges are 2mm apart. The cut of the created grate was evaluated according to a Gt0–Gt4 scale.

2.5.5 Overall evaluation of the selected physical–mechanical properties of the paints

The measured quantities of physical-mechanical nature including the adhesion of the paint film identified by means of a cross-cut test, impact resistance, and resistance against cupping indicate the elasticity, elongation at break, and strength of the paint film. The detected results of the mechanical tests were assigned with the corresponding numerical values from the scale for the determination of physical-mechanical properties. The high value of the different mechanical properties means that the pigment contained in the paint has a positive influence on its mechanical properties.

2.5.6 The effect of pigment particles on the surface hardness of the paints

Good surface resistance is important for the resistance of the paints against mechanical impacts, scratches, erosion caused by abrasion of dust particles in the outside atmospheric environment. The determination of the surface hardness of the paint film on glass was carried out by means of a pendulum apparatus, the test consists of measuring the number of oscillations of the pendulum that bears onto the paint film with two steel balls. The unit of measuring of hardness is a percentile value related to the hardness of a glass standard that equals to 100%.

2.6 The effect of the prepared pigments on the anticorrosion performance and the chemical resistance of the paints

The primary goal of this study was to prepare a pigment that would increase the anticorrosion performance of the paints. The determination of the properties of the prepared pigments by means of laboratory tests and electrochemical measurements in corrosive environments was a priority assignment.

2.6.1 Corrosion test evaluation methods

The evaluation of coatings after exposure to the corrosion tests followed methods based on the (ASTM D 714-87) for degree of blistering, (ASTM D 6294-98) for degree of rusting, and (ASTM D 2803-93) for photographic inspection.

2.6.2 Electrochemical evaluation method

Electrochemical impedance spectroscopy (EIS) is an extremely useful technique in generating quantitative data that relates to the quality of the coat on a metal substrate. EIS is a very sensitive detector of a coated metal condition, its response can indicate changes in the coating long before any visible damage occurs, and it is not an absolute measurement.

EIS is useful to characterize a painted metal substrate by simultaneously measuring two phenomena: (1) the deterioration of the organic coating caused by exposure to an electrolyte and (2) the increase in corrosion rate of the underlying substrate due to the deterioration of the coating and subsequent attack by the electrolyte.

In EIS, an AC voltage of varying frequency is applied to the sample. It is useful to think of the frequency as a camera shutter that can be very fast (high frequency) for fast reactions and very slow (low frequency) for slow ones. This is the technical feature that allows EIS to gather so much information on an electrochemical reaction in one experiment, and this is why EIS is more useful for coatings than DC electrochemical techniques.

EIS can quantitatively measure both resistances and capacitances in the electrochemical cell. Resistance corresponds to electron transfer reactions such as corrosion, while the capacitance of a metal electrode in contact with an electrolyte is important information for any electrochemical system, for organic coatings, the capacitance measurement is particularly revealing. As the organic coating deteriorates with time during exposure to an electrolyte, EIS can track changes in the capacitance of the coating. The capacitance will change as the coating swells or absorbs water. In addition, changes in the porosity of the coating can be easily measured. EIS can also simultaneously monitor the rate of corrosion of the metallic substrate which generally increases as the protective coating fails, allowing the electrolyte to contact the substrate (Mirabedini et al, 2003; Kendig,et al., 1983; Hernandez et al., 1999; Huang et al., 2008).

In the present work, EIS experiments were carried out using a conventional three-electrode cell. The working electrode was the coated specimen, using a saturated calomel electrode as reference electrode, and a platinum foil (1 cm^2) as counter electrode. The used electrolyte was 3.5 wt. % NaCl solution. EIS measurements were carried out using AC signals of amplitude 5mV peak to peak at the open circuit potential in the frequency range between 15 kHz and 0.3 Hz. EIS data was collected using Gamry PCI300/4 Potentiostat/Galvanostat/Zra analyzer, EIS300 Electrochemical Impedance Spectroscopy software, and Echem Analyst 5.21 for results plotting, graphing, data fitting & calculating.

3. Result and discussion

3.1 The physical–chemical properties of the prepared pigments

Silica fume is a fine amorphous kind of silica, while titanium dioxide possess well crystalline particles, since the layers of titanium dioxide were so thin that they do not exceed 10% of the whole pigment concentration as can be seen in Table (3), they do not alter the crystals of silica and hence the XRD charts do not show any difference between the amorphous structure of uncovered silica fume and the silicas covered with the three concentrations of titanium dioxide. This can be clearly seen in Figure (1).

Figures (2 and 3) show the morphology of the prepared pigments using SEM and TEM. From the featured photos it can be seen that, silica fume possess a spherical particle shapes, while titanium dioxide possess platelet structure. The new pigments possess platelet

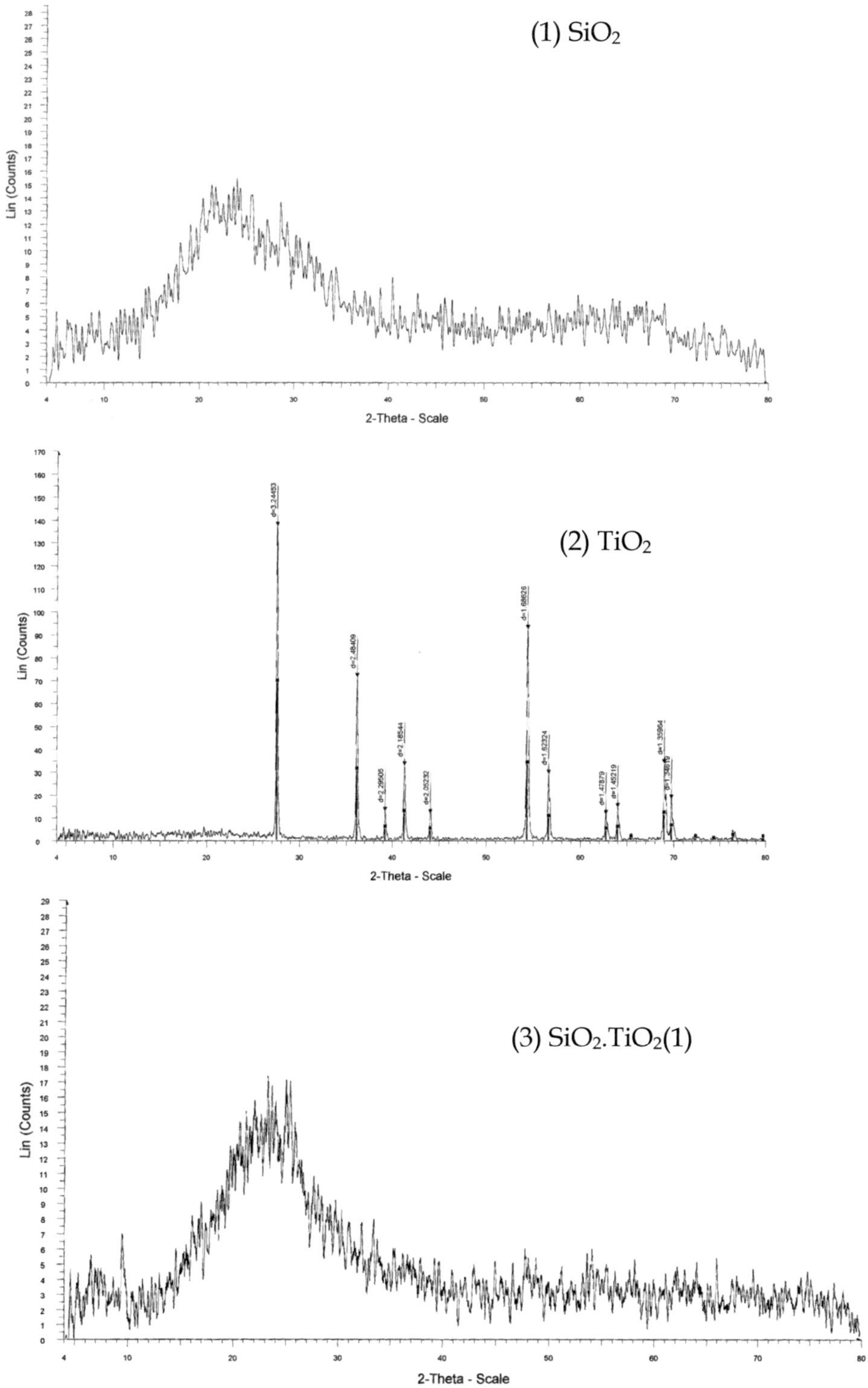
(1) SiO_2
Lin (Counts)
2-Theta - Scale
(2) TiO_2
Lin (Counts)
2-Theta - Scale
(3) $SiO_2.TiO_2(1)$
Lin (Counts)
2-Theta - Scale

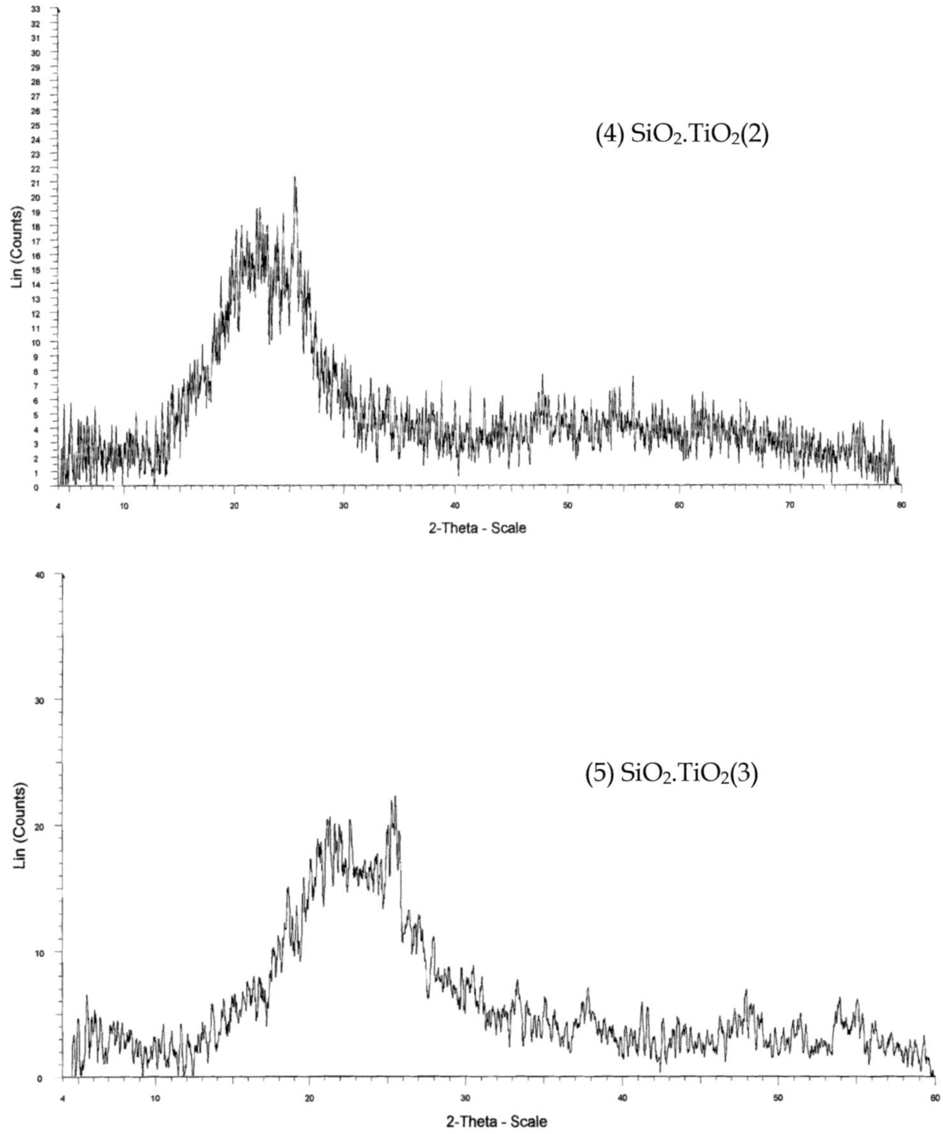

Fig. 1. XRD charts of 1. silica fume, 2. Titanium dioxide, 3. Silica coated with titanium dioxide (1), 4. Silica coated with titanium dioxide (2), and 5. Silica coated with titanium dioxide (3)

1. Silica fume

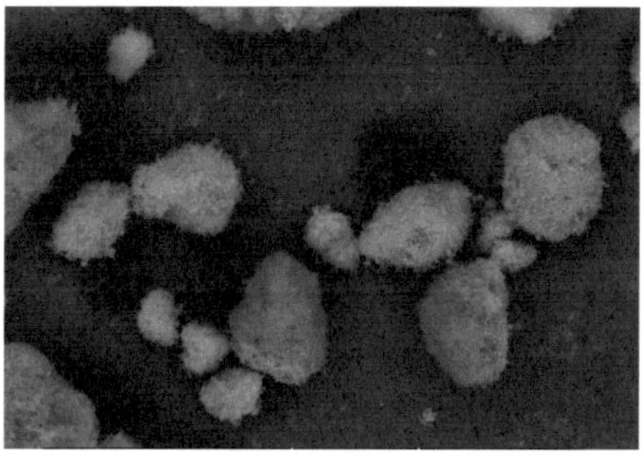

2. Titanium dioxide

3. Silica covered with TiO_2 (1)

4. Silica covered with TiO_2 (2)

5. Silica covered with TiO_2 (3)

Fig. 2. SEM micrographs of 1.Silica, 2.Titanium dioxide, 3, 4 and 5.Silica covered

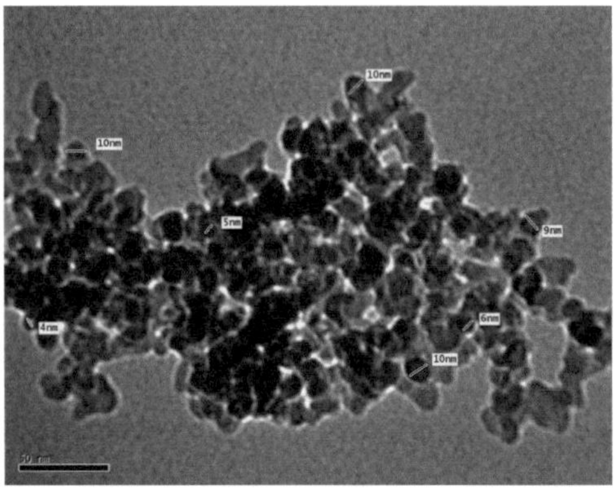

1. Silica fume

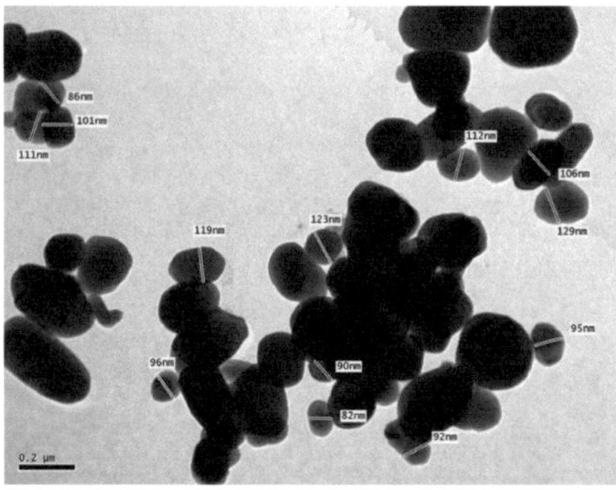

2. Titanium dioxide

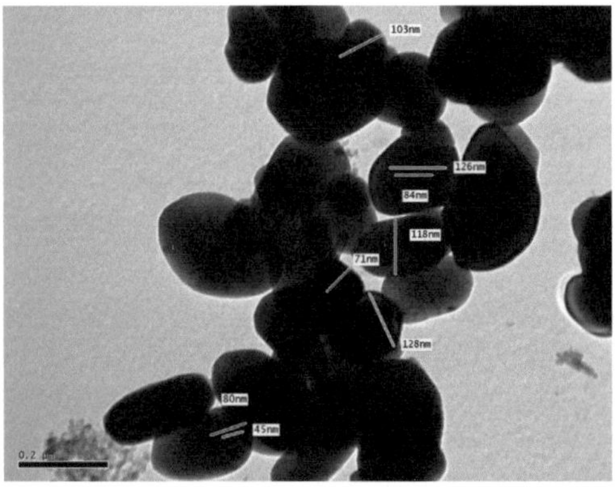

3. Silica covered with TiO_2 (1)

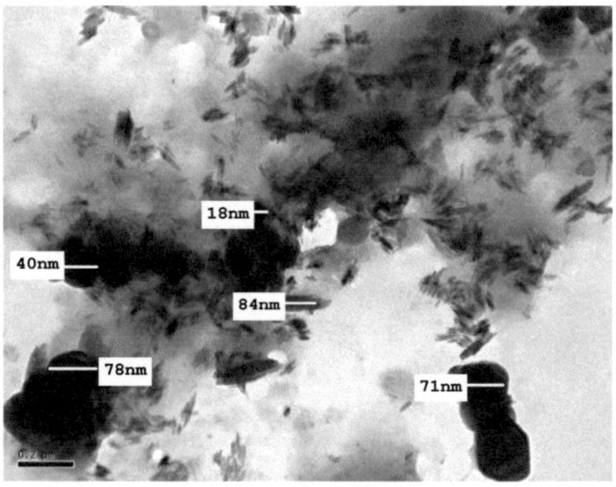

4. Silica covered with TiO_2 (2)

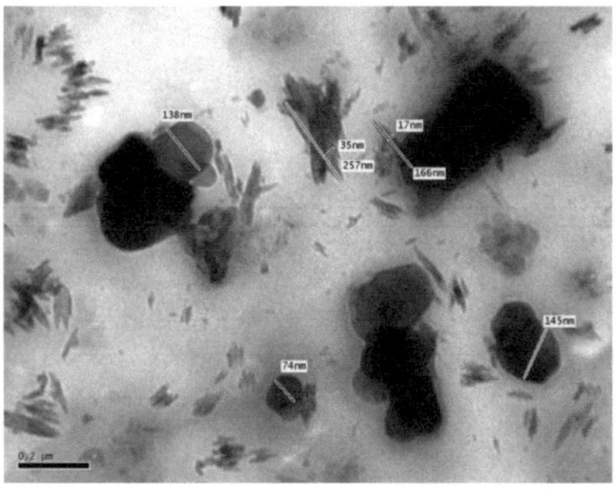

5. Silica covered with TiO_2 (3)

Fig. 3. TEM micrographs of 1. Silica, 2. Titanium dioxide, 3, 4 and 5 Silica covered with different concentrations of Titanium dioxide at magnification 120X

structures but of different sizes, this can be obviously seen in the TEM photos shown in Figure (3), where the silica spheres were clearly shown in Figure 3a, while plates of titanium dioxide are shown in figure 3b, and as the titanium begin to deposit on silica surface, new structures begin to appear with the silica appearing as clouds in the background of the photo due to their purity and fine particles with rods or small plates of titanium depositing on its surface with their amounts increasing from Figure 3d to Figure 3e. These plates provide a reinforcing effect reducing the water and gas permeability, and therefore imparting good anticorrosive properties and special appearance to the paint film when used in paint formulations (Lambourne, 1987).

Figure (4) shows the EDAX analysis of prepared pigments. EDAX or energy dispersive X-ray analysis can detect the elements on the surface up to one micron depth. As featured from the chart, titanium was detected; this revealed its presence on silica surfaces.

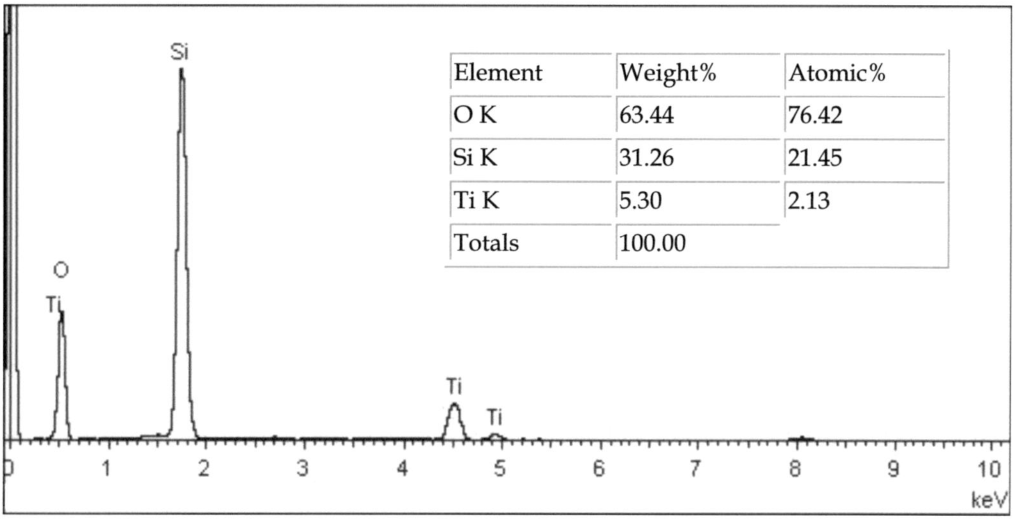

Fig. 4. EDAX analysis of silica fume covered with titanium dioxide

Table 3. represents the XRF analyses results of the three prepared pigments besides the uncovered silica fume. From the Table it can be seen that titanium concentration differs in an ascending order from SiO_2/TiO_2 (1) to SiO_2/TiO_2 (3), revealing that different concentrations of

Concentration of main constituents (wt. %)	Silica fume	TiO_2/SiO_2 (1)	TiO_2/SiO_2 (2)	TiO_2/SiO_2 (3)
SiO_2	95.54	94.48	93.61	91.78
TiO_2	----	2.85	3.90	5.72
Al_2O_3	0.34	0.36	0.37	0.35
Fe_2O_3	0.91	1.43	1.27	1.23
MgO	0.70	0.60	0.60	0.57
CaO	0.32	0.31	0.33	0.32
Na_2O	0.39	0.18	0.17	0.18
K_2O	0.80	0.95	0.98	0.86
P_2O_5	0.07	0.06	0.06	0.06
SO_3	0.005	0.005	0.005	0.007
L.O.I	4.93	0.62	0.53	0.76

Table 3. XRF analysis results of the prepared TiO_2/Silica fume pigments Specifications of prepared SiO_2/TiO_2 pigment properties

titanium dioxide is present on surface of silica and these concentrations do not exceed 10% of the whole pigment concentration.

Specifications of prepared TiO_2/SiO_2 pigment properties

From Table (4) it is clear that;

a. Specific gravity of silica was null, it is a very light fluffy powder, but the prepared TiO_2/ SiO_2 pigments have specific gravity values which are in direct relation with the concentration of titanium dioxide on silica surface.
b. Oil absorption of silica is very high and this is expected because of the silica spherical shapes which allow more oil to be deposited in the voids between the particles to form a paste, but TiO_2/ SiO_2 pigments have lower oil absorption than silica. As it is well known that oil absorption expresses how much binder can be consumed by the pigment, and thus how this pigment is economically feasible when used in paint formulations. As the oil absorption is higher, more binder will be needed to completely wet the pigment and form a homogeneous paint film.
c. Bulking value (BV) expresses the volume of paint for a given weight of a pigment, therefore pigments with higher bulking values as in silica and TiO_2/ SiO_2 can be considered more economic.
d. pH values of the different pigments are neutral.

In general, the new pigments possess better physical properties than either of its components individually. This was obvious from the specific gravity and oil absorption of these core-shell pigments.

Materials	**Oil absorption (g/100g)**	**Sp.Gr.**	**BV (gal/100b)**	**pH**	**Whiteness (%)**
SiO_2	819.68	0	0	7	97
TiO_2	22	4.24	2.842	7	95
TiO_2/ SiO_2 (1)	221.84	1.59	7.578	7	97
TiO_2/ SiO_2 (2)	193.64	1.53	7.875	7	98
TiO_2/ SiO_2 (3)	189.88	1.78	6.770	7	98

Table 4. Characteristics of prepared titanium dioxide/kaolin pigments

3.2 The effect of prepared pigments on mechanical properties of paints

Tables (5-7) represent the paint formulations, while Table (8) expresses the results of determining hardness of organic coatings by means of pendulum apparatus; also impact and ductility were determined. Paint films containing silica show the best mechanical properties among the group, while those containing titanium dioxide show the lowest values, this is may be due to the spherical particles of silica which gives the film elasticity due to the voids present between these spheres. This can be in accordance with the properties of silica which are almost in the nano-scale and paint films containing nano-pigments have the following properties;

Constituent (%wt.)	1	2	3	4	5
Fe_2O_3 (Hematite)	10	10	10	10	10
Kaolin	15	15	15	15	15
ZnO	7	7	7	7	7
SiO_2	30	---	---	---	---
TiO_2	---	30	---	---	---
TiO_2/SiO_2 (1)	---	---	30	---	---
TiO_2/SiO_2 (2)	---	---	---	30	---
TiO_2/SiO_2 (3)	---	---	---	---	30
Total pigment	63.5	63.5	63.5	63.5	63.5
Total binder	36.5	36.5	36.5	36.5	36.5
P/B	1.74	1.74	1.74	1.74	1.74
Wetting & dispersing agent	1	1	1	1	1
Drier	0.5	0.5	0.5	0.5	0.5
Total	100	100	100	100	100

Table 5. Paint formulations of TiO_2/SiO_2 pigments with medium oil alkyd resin (Group I)

Constituent (%wt.)	6	7	8	9	10
Fe_2O_3 (Hematite)	15	15	15	15	15
Kaolin	15	15	15	15	15
ZnO	7	7	7	7	7
SiO_2	30	---	---	---	---
TiO_2	---	30	---	---	---
TiO_2/SiO_2 (1)	---	---	30	---	---
TiO_2/SiO_2 (2)	---	---	---	30	---
TiO_2/SiO_2 (3)	---	---	---	---	30
Total pigment	68.5	68.5	68.5	68.5	68.5
Total binder	31.5	31.5	31.5	31.5	31.5
P/B	2.175	2.175	2.175	2.175	2.175
Wetting & dispersing agent	1	1	1	1	1
Drier	0.5	0.5	0.5	0.5	0.5
Total	100	100	100	100	100

Table 6. Paint formulations of TiO_2/SiO_2 pigments with medium oil alkyd resin (Group II)

According to (Guin et al., 2011; Otterstedt & Brandreth, 1998; Salamone, 1999), the main advantages of nano-coating or coatings containing nano-pigments are:

- Better surface appearance.
- Good chemical resistance.
- Decrease in permeability to corrosive environment and hence better corrosion properties.
- Optical clarity.
- Increase in modulus and thermal stability.

- Easy to clean surface.
- Anti-skid, anti-fogging, anti-fouling and anti-graffiti properties.
- Better thermal and electrical conductivity.
- Better retention of gloss and other mechanical properties like scratch resistance.
- Anti-reflective in nature
- Chromate and lead free
- Good adherence on different type of materials.

In general, TiO_2/SiO_2 pigments show less mechanical properties than silica, but better than paint films containing titanium dioxide and this is due to that as titanium dioxide layers precipitated on silica surface transfers the particle sizes to the micron-scale and also they disturb the texture of silica by altering their platelet particles between the silica spheres leading to less homogenous texture and thus less mechanical properties. Paint films containing titanium dioxide exhibit poor mechanical properties.

Constituent (%wt.)	11	12	13	14	15
Fe_2O_3 (Hematite)	21.5	21.5	21.5	21.5	21.5
Kaolin	15	15	15	15	15
ZnO	7	7	7	7	7
SiO_2	30	---	---	---	---
TiO_2	---	30	---	---	---
TiO_2/SiO_2 (1)	---	---	30	---	---
TiO_2/SiO_2 (2)	---	---	---	30	---
TiO_2/SiO_2 (3)	---	---	---	---	30
Total pigment	75	75	75	75	75
Total binder	25	25	25	25	25
P/B	3	3	3	3	3
Wetting & dispersing agent	1	1	1	1	1
Drier	0.5	0.5	0.5	0.5	0.5
Total	100	100	100	100	100

Table 7. Paint formulations of TiO_2/SiO_2 pigments with medium oil alkyd resin (Group III)

3.3 The effect of prepared pigments on anticorrosive properties of paints

Table (8) features the results of determining the anticorrosion performance of paint films through blistering on paint surface and rust under film. The method classifies the osmotic blisters to the groups according to their sizes designated by numbers 2, 4, 6, and 8 (2 denotes the largest size, 8 the smallest one).

To the blister size information on the frequency of occurrence is given. The highest occurrence density of blisters is designated as D (dense), the lower ones as MD (medium dense), M (medium) and F (few). In such away a series from the surface area attacked at least by the osmotic blisters up to the heaviest occurrence can be formed as follows: 8F-6F-4F-2F-8M-6M-4M-2M-8MD-6MD-4MD-2MD-8D-6D-4D-2D.

Group I					
Drying time, *hr.*	**1**	**2**	**3**	**4**	**5**
Surface dry			1 - 2		
Thorough dry			3 – 4		
Adhesion	Gt0	Gt0	Gt0	Gt0	Gt0
Hardness, *Sec (90μ)*	60	67	74	72	73
Ductility, *Mm*	6.3	3	5.3	5.1	5.1
Impact resistance, *Kg.m.*	0.75	0.325	0.6	0.563	0.525
Corrosion resistance					
Degree of Blistering	10	6M	10	10	10
Degree of rusting	8-G	7-S	7-P	6-G	5-G
Filiform corrosion			Fig.1-FA		
Group II					
Drying time, *hr.*	**6**	**7**	**8**	**9**	**10**
Surface dry			1 - 2		
Thorough dry			3 – 4		
Adhesion	Gt1	Gt0	Gt1	Gt1	Gt1
Hardness, *Sec (90μ)*	72	73	78	80	82
Ductility, *Mm*	6.7	2.7	5.8	5.5	5.4
Impact resistance, *Kg.m.*	0.738	0.275	0.55	0.488	0.45
Corrosion resistance					
Degree of Blistering	4MD	8M	8F	10	4MD
Degree of rusting	5-G	10	7-P	10	5G
Filiform corrosion			Fig.1-FA		
Group III					
Drying time, *hr.*	**11**	**12**	**13**	**14**	**15**
Surface dry			1 - 2		
Thorough dry			3 – 4		
Adhesion	Gt1	Gt0	Gt1	Gt1	Gt1
Hardness, *Sec (90μ)*	78	82	88	92	92
Ductility, *Mm*	6.4	4.3	5.2	5.2	5.0
Impact resistance, *Kg.m.*	0.738	0.263	0.525	0.488	0.425
Corrosion resistance					
Degree of Blistering	4MD	4M	6M	6MD	4MD
Degree of rusting	7-S	10	8-P	6-S	4-S
Filiform corrosion			Fig.1-FA		

8-G, 0.1% general rusting; 7-S, 0.3% spot rusting; 7-P, 0.3% pinpoint rusting; 6-G, 1% general rusting; 5-G, 3% general rusting; 8-P, 0.1% pinpoint rusting, 6-S, 1% spot rusting, and 4-S, 10% spot rusting.

Table 8. Physico-mechanical and corrosion characteristics of dry paint films

As can be detected from Tables (5-8) and Figures (5-7), that paint films of group II were better in their performance than the other groups in the rust under paint films revealing that the P/B 2.175 is better than the other P/B ratios. Also paint films containing SiO_2/TiO_2 (1) and SiO_2/TiO_2 (2) were better in their performance than SiO_2/TiO_2 (3), this may be due to

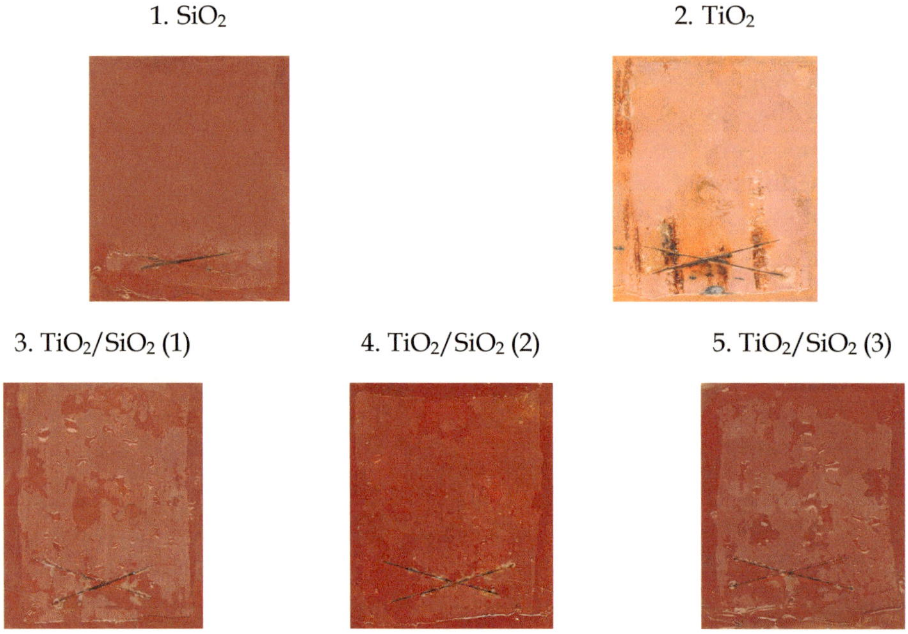

Fig. 5. Photos of paint films of group I after immersion in 3.5% NaCl after 28 days

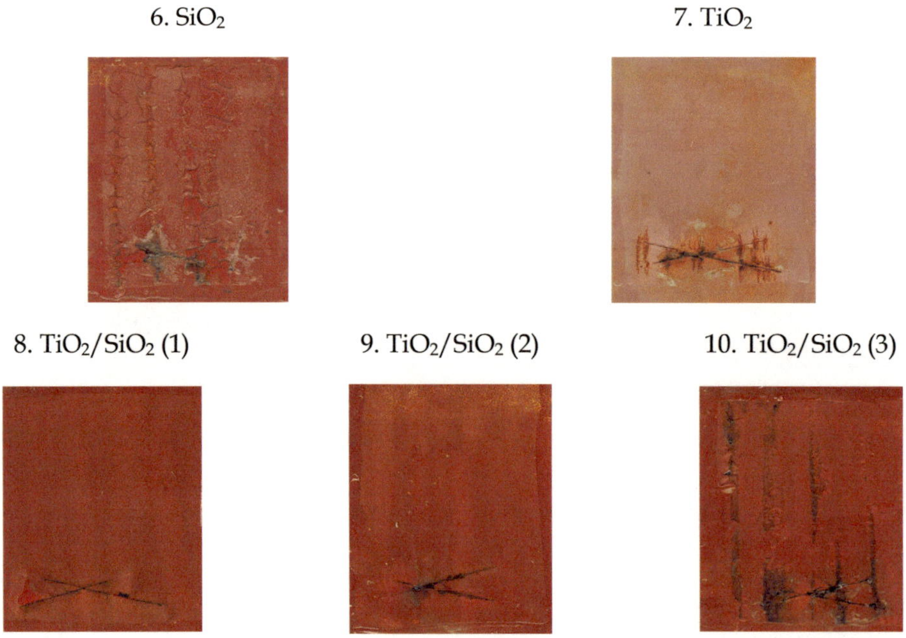

Fig. 6. Photos of paint films of group II after immersion in 3.5% NaCl after 28 days

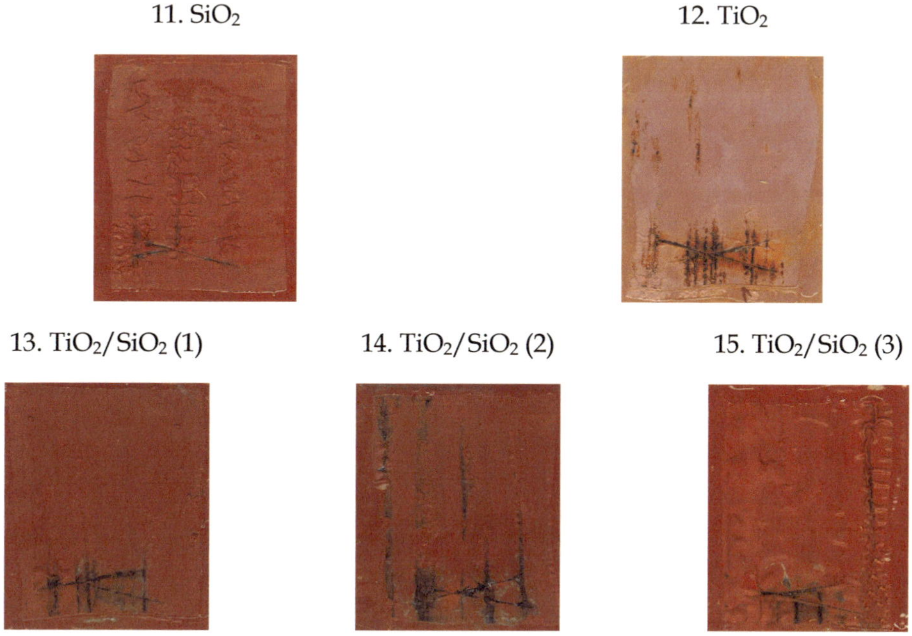

Fig. 7. Photos of paint films of group III after immersion in 3.5% NaCl after 28 days

the photocatalytic effect of titanium dioxide on the surface of silica which is in accordance with its concentration, i.e. as titanium dioxide concentration increases, the photocatalytic effect also increases. This behavior can be also explained according to that as concentration of titanium increases, it alters the silica particles causing random effect between the particles and hence less film compactness leading to easiness of the passage of corroding materials to the metal substrate.

3.4 Electrochemical studies on paint films

EIS was employed to investigate the corrosion protection performance TiO_2/SiO_2 core-shell pigments in anticorrosive paint formulations based on different P/B ratios on the protection of mild steel substrate in 3.5 wt% NaCl at different immersion times. The results of the analysis and calculations of the impedance data, at different immersion times 1, 7, 14, 21 and 28 days, were listed in Table 9.

Inspections of this data indicate the following;

a. The coating resistance (*Rc)* is clearly higher in case of group II (P/B 2.175) paint formulations than that of other groups at all immersion times. It is obvious that the rate of corrosion decrease in the order: group II > group I > group III.
b. The coating resistance (*Rc)* of paint films containing TiO_2/SiO_2 (1) and TiO_2/SiO_2 (2) were better in their performance than SiO_2/TiO_2(3).
c. As shown in Figure 5, the paint film containing TiO_2/SiO_2(1) in group II (paint no. 8) offered a very good resistance even after 28 days immersion in 3.5 wt% NaCl solution (9.809 K ohm), indicating that paint formulations containing TiO_2/ SiO_2 (1) in P/B 2.175

is the best among the three groups and they can provide effective protection to carbon steel.

d. The coating resistance (*Rc*) is inversely proportional to immersion times. As paint no. 8 was the best among the groups it was taken as an example to show in Figure 5 the typical Nyquist plots after different immersion time in 3.5 wt% NaCl.

Group	No.	Sample	R_c (Kohm.cm^2)				
			1 day	7 day	14 day	21day	28 day
I	1	SiO_2	2.197	1.114	0.692	0.541	0.474
	3	$TiO_2/SiO_2(1)$	3.271	2.845	2.696	2.388	1.174
	4	$TiO_2/SiO_2(2)$	3.923	0.604	0.430	0.374	0.265
	5	$TiO_2/SiO_2(3)$	1.038	0.764	0.681	623.7	0.503
II	6	SiO_2	0.933	0.660	0.658	0.372	0.271
	8	$TiO_2/SiO_2(1)$	147.34	25.36	13.09	11.63	9.809
	9	$TiO_2/SiO_2(2)$	62.67	9.896	5.693	4.012	3.033
	10	$TiO_2/SiO_2(3)$	2.015	1.192	0.790	0.564	0.486
III	11	SiO_2	0.145	0.135	0.109	0.082	0.042
	13	$TiO_2/SiO_2(1)$	1.158	0.700	0.598	0.401	0.355
	14	$TiO_2/SiO_2(2)$	1.021	0.513	0.489	0.388	0.329
	15	$TiO_2/SiO_2(3)$	0.331	0.105	0.079	0.060	0.056

Table 9. EIS results for different paint formulations in 3.5 wt% NaCl at different immersion times

From the Figure it can be detected that the decrease in coating resistance may be due to the penetration of water and movement of ionic species among the coating layer, increasing the coating conductivity. This can be explained according to the following steps;

- Initially, the electrolyte penetrates through the coating layer, and sets up conducting paths at different depths within the coating (Xianming et al., 2009). With increase immersion time, the electrochemical reactions at the interface between the coating and the metal surface make progress where the electrolyte phase meets the metal/oxide interface and a corrosion cell is then activated.
- This step is Followed by that the barrier properties of the coating are decreased, suggesting a decrease in the coating resistance, i.e. decrease the radius of the semi-circle (Del Amo et al., 1999).

Generally, electrochemical studies were in high accordance with accelerated laboratory test results.

3.4.1 Some suggested mechanisms of protection concerning the new prepared pigments

- Silica is used in coatings and paints to increase the scratch resistance. The silica fine particles do not interfere with the visible light, only by a high quality of dispersion, cloudiness can be avoided and transparency can be achieved. The high tendency of silica particles to agglomerate with each other makes it difficult to assure that every single silica particle can react with the surrounding medium. The reaction of the individual particles with the medium is important for many applications in order to achieve the desired results (Kirubaharan et al., 2009).

- Hydrophobic silica fume is used to enhance water and corrosion resistance this is due to improved pigment stabilization in the coating system combined with the stronger barrier effect inherent to hydrophobic silica fume. This barrier effect in particular is what prevents moisture from penetrating the coating film and reaching the substrate (Guin et al., 2011).

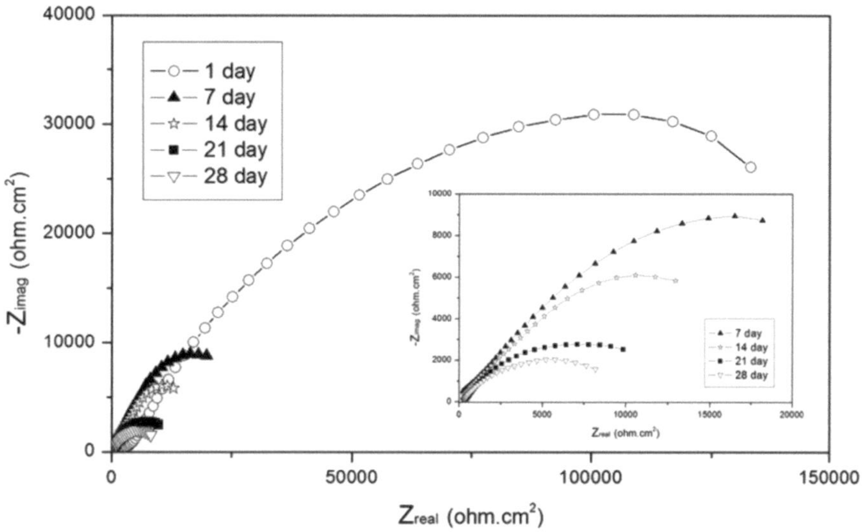

Fig. 9. Nyquist plots for TiO_2/SiO_2(1) of group II (paint no. 8) in 3.5 wt% NaCl at different immersion times 1, 7, 14, 21 and 28 days

- Fine Silica fume is used as additive that enhance the surface smoothness and dispersability of the media into which they are added. Their fine silica particles have a lower coefficient of friction than titanium dioxide and other popular fillers. On a molecular level, these particles are perfectly spherical in shape and move more freely to provide a superior tactile feel which serve as a matting agent for paints besides being corrosion resistant (Kirubaharan et al., 2009).
- Durability is an important property of titanium dioxide, and since it is present in only very low concentrations in the prepared pigments, its opacity is controlled, but in spite of this it help in increasing the hiding power and durability of the new prepared pigments (Salamone 1999; Patton 1973; Ahmed & Selim, 2010).
- Titanium dioxide can also contribute to some extent to the preservation of the physical integrity of the medium in which it is dispersed by providing protection against cracking, checking, loss of adhesion and loss of tensile strength (Ahmed & Selim 2010; Ahmed & Selim, 2011).
- Titanium dioxide particles forming the shell were arranged in alignment between the silica particles comprising the core. In case of low concentration of titanium dioxide, its

small platelet particles order themselves in between the voids of the spherical silica particles in a close-pack texture locking the voids between silica particles and thus more compact paint film will be formed, prohibiting the formation of blisters or rust under film. As the concentration of titanium dioxide increases, a less compact film will be formed and a disturbance between silica particles occurs, these spaces formed between the binder and the pigment particles, even under the best circumstances give the chance for areas arise on the surface of the pigment particle where the binder and the particle may be in extremely close physical proximity but are not chemically bonded. This area between binder and pigment can be a potential route for water molecules to slip through the cured film (Forsgren, 2006, Ahmed & Selim, 2011).

- Titanium dioxide is an inert oxide which imparts barrier properties to organic coatings by impeding the transport of aggressive species to the surface of the substrate. Such pigments orientate themselves parallel to the substrate surface and protect the substrate by providing a tortuous path of diffusion to the substrate. In addition, they may have a reinforcing effect on the mechanical properties of the coating. Its platy particles may be one reason of its protection properties; also it gives excellent hiding power to the films. Titanium dioxide is of high cost and its presence in high concentration may lead to the degradation of paint films due to its photochemical reactivity (Sorensen, 2009, Forsgren, 2006; Ahmed & Selim, 2010; Ahmed & Selim, 2011).
- As concentration of titanium dioxide is low, there is no chance for its photochemical reactivity to show itself (Bremmell & Mensah, 2005).

4. Conclusions

1. New pigments based on core-shell method were prepared in this work; these pigments are based on silica fume that is an industrial byproduct which was covered with different concentrations of titanium dioxide.
2. The pigments overcome the low hiding power of silica fume, and the new pigments possess better hiding power that was directly related to the titanium dioxide concentration on the silica surface.
3. The corrosion protection performance was inversely related to the concentration of titanium dioxide, this was because its platy particles alters the spherical silica particles leading to disturbing the texture that leads to less homogenous film and thus less protection.
4. The best corrosion performance among the pigments was TiO_2/SiO_2 (1) and the best among the groups was group II with P.B 2.175.
5. Electrochemical studies were in high agreement with the accelerated laboratory test.

5. Acknowledgment

This work was supported by Science and Technology Development Fund (STDF) under project number 1242.

6. References

Abu-Ayana, Y.M., Yossef, E.A.M. and El-Sawy, S.M. (2005). Silica fume - formed during the manufacture of ferrosilicon alloys - as an extender pigment in anticorrosive paints, *Anti-Corrosion Methods and Materials* Vol. 52, No. 3, pp.345 - 352.

Ahmed, N.M. and Selim, M.M. (2010). Anticorrosive performance of titanium dioxide-talc hybrid pigments in alkyd paint formulations for protection of steel structures, *Anticorrosion Methods and Materials* Vol. 57, No. 1, pp. 133-141.

Ahmed, N.M. and Selim, M.M. (2011). Innovative titanium dioxide-kaolin mixed pigments performance in anticorrosive paints, *Pigment and Resin Technology* Vol. 40, No. 1, pp. 4-16.

Barnett, J.R., Miller, S., and Pearce E. (2006). Colour and art: A brief history of pigments. *Optics & Laser Technology* Vol. 38, No. 1, pp. 445–453.

Bremmell, K.E. and Mensah, K.E. (2005). Interfacial chemistry mediated behavior of colloidal talc dispersions, *J. of Colloid & Interface Sci.* Vol. 283, No. 1, pp. 385-391.

Brock, T. (2000). European Coatings Handbook, p. 157, Elsevier.

Brock, T. Groteklaes, M. and Mischke, P. (2000). European coatings handbook, p. 45, Elsevier.

Buxbaum, G. and Pfaff, G. 2005. Industrial inorganic pigments, Chap. 2, p.201, Wiley interscience, N.Y.

Degryse, P. Elsen, J. (2003). Industrial minerals: resources, characteristics, and applications, pp. 320-323, Leuvin University Press, Netherlands.

Del Amo, B. Romagnoli, R. Deya, C. and Gonzalez, J.A. (2002). High performance water-based paints with non-toxic anticorrosive pigments, *Progress in Organic Coatings* Vol. 45, No. 3, pp. 389-397.

Del Amo, B. Romagnoli, R. and Vetere, V.F. (1999). Steel Corrosion Protection by Means of Alkyd Paints Pigmented with Calcium Acid Phosphate, *Ind. Eng. Chem. Res.* Vol. 38, No. 3, pp. 2310-2314.

Forsgren, A. (2006). Corrosion Control through Organic Coatings, (Chap. 2), pp. 37-39, Taylor & Francis.

Gaumet, S. Siampigingue, N. Lemaire, J. and Pacaud, B. (1997). Influence of titanium dioxide pigment characteristics on durability of four paints (acrylic isocyanate, polyester melamine, polyester isocyanate, and alkyd), *Surface Coatings International Part B: Coatings Transactions* Vol. 80, No.2, pp.367-372.

Guin, A. K. Nayak, S. K. Rout, T. K. Bandyopadhyay, N. Sengupta, D. K. (2011). Corrosion behavior of nanohybrid titania–silica composite coating on phosphated steel sheet, *J. Coat. Technol. Res.*, online since 24 Feb.

Gysau, D. (2006). Fillers for paints: fundamentals and applications, p. 32, European coatings literature.

Herbst, W. and Hunger, K. (2006). Industrial organic pigments, Chap. 1, p. 143, Wiley VCH, GmBH.

Hernández, L.S. Del Amo, B. and Romagnoli, R. (1999). Accelerated and EIS tests for anticorrosive paints pigmented with ecological pigments, *Anti-Corrosion Methods and Materials* Vol. 4, No. 2, pp. 198–204.

Huang, Y. Shih, H. Huang, H. Daugherty, J. Wu, S., Ramanathan, S. Chang, C., and Mansfeld, F. (2008). Evaluation of the corrosion resistance of anodized aluminum 6061 using electrochemical impedance spectroscopy (EIS), *Corrosion Science* Vol. 50, No. 5, pp. 3569–3575.

Kendig, M. Mansfeld, F. Tsai, S. (1983). Determination of the Long Term Corrosion Behavior of Coated Steel with A.C. Impedance Measurements, *Corrosion Science* Vol. 23, No. 3, pp. 317–329.

Kirubaharan, A. M. K. Selvaraj, M. Maruthan, K. Jeyakumar, D. (2009). Synthesis and characterization of nanosized titanium dioxide and silicon dioxide for corrosion resistance applications, *J. Coat. Technol. Res.* Vol. 21, online since 25, Nov.

Lambourne, R. (1987). Paint and surface coatings, p. 131, Ellis Horwood Ltd. Pub.

Martens, C.R. (1974). Technology of Paints, Varnishes, and Lacquers, p. 344, Robert. E. Krieger Publishing Co., Melbourne, FL.

Mirabedini, S.M. Thompson, G.E. Moradian, S. Scantlebury, J.D. (2003). Corrosion performance of powder coated aluminium using EIS, *Progress in Organic Coatings* Vol. 46, No. 2, pp. 112–120.

Molera, P. Oller, X. Del Valle, M. and Gonzalez, F. (2004). Formulation and characterization of anticorrosive paints, *Pigment & Resin Technology* Vol. 33. No. 1, pp.99–104.

Otterstedt, J.E. Brandreth, D.A. (1998). Small particles technology, Chap. 4, pp. 176-178, Technology and Engineering.

Patton, T.C. (1973). Pigment handbook, p. 199, Wiley-Interscience publication.

Perera, D.Y. (2004). Effect of pigmentation on organic coating characteristics, (review), *Progress in Organic Coatings* Vol. 50, No. 1, pp.247-262.

Peter, H. and Robert, V. (1999). High grade kaolin fillers–production review: Industrial Minerals, Vol. 54, pp 25-37.

Salamone, J.C. (1999). Concise polymeric materials encyclopedia, p. 477, CRC press.

Shao, Y. Jia, C. Meng, G. Zhang, T. Wang, F. (2009). The role of a zinc phosphate pigment in the corrosion of scratched epoxy-coated steel, *Corrosion Science* Vol. 51, No. 1, pp. 371–379.

Siddique, R. and Khan, M.I. (2011). Supplementary Cementing Materials, Chapter 2, pp. 67-117, Engineering Materials, Springer-Verlag Berlin Heidelberg.

Sorensen, P.A., Kiil, S. Dam-Johansen, K. and Weinell, C.E. (2009). Anticorrosive coatings: a review, *J. Coat. Technol. Res.* Vol. 6, No. 1, pp.135–176.

Streitberger, H.J. and Kreis, W. (2008). Automotive paints and coatings, p. 144, Science and Technology publishing.

Talbert, R. (2007). Paint technology handbook, p. 61, Taylor & Francis Inc.

Veleva, L., Chin, J. and Del Amo, B. (1999). Corrosion electrochemical behavior of epoxy anticorrosive paints based on zinc molybdenum phosphate and zinc oxide, *Progress in Organic Coatings* Vol. 36, No. 2, pp. 211-216.

Vesely, D., Kalendova, A. (2008). Anticorrosion efficiency of $Zn_xMg_yAl_2O_4$ core–shell spinels in organic coatings, *Progress in Organic Coatings* Vol. 62, No.1, pp. 5–20.

Walter, G.W. (1991). Laboratory Simulation of Atmospheric Corrosion by SO_2-II. Electrochemical Mass Loss Comparisons, *Corrosion Science* Vol.32, No. 5, pp. 1353–1359.

Warson, H., Finch, C.A. (2001). Applications of synthetic resin lattices: Lattices in surface coatings, p. 172, Technology and Engineering.

Weil, T.C. (2011). Addressing Parking Garage Corrosion with silica Fume, Transportation Research Record 1204, pp. 8-10, W. R. Grace & Company, Construction Products Division, Cambridge, Mass.

Xianming, S. Nguyen, T.A. Zhiyong, S. Liu, Y. and Avci, R. (2009). Effect of nanoparticles on the anticorrosion and mechanical properties of epoxy coating, *Surface & Coatings Technology* Vol. 204, No.1, pp. 237–245.

Yingchao, Z. Hongqi, Y. Hui, L. Kai, H. (2011). Preparation and characterization of aluminium pigments coated with silica for corrosion protection, *Corrosion Science*, article in press.

5

Improvement of the Corrosion Resistance of Carbon Steel by Plasma Deposited Thin Films

Rita C.C. Rangel, Tagliani C. Pompeu, José Luiz S. Barros Jr., César A. Antonio, Nazir M. Santos, Bianca O. Pelici, Célia M.A. Freire, Nilson C. Cruz and Elidiane C. Rangel
Paulista State University,
University of Campinas,
Brazil

1. Introduction

It is estimated that nearly 3% of the global domestic gross product, corresponding to 2.8 trillion US dollars, is wasted every year with problems related to corrosion (Koch et al., 2002). Oil and gas companies, for instance, spend up to US$ 80 billion only with the corrosion of devices in marine environments (Muntasser et al., 2002). About 90% of the corrosion costs are associated with iron-based materials. Since carbon steels, which account for about 85% of the annual worldwide steel production, represent the largest class of iron-alloys in use, the corrosion of such materials is of paramount importance. Notwithstanding many years of intensive research and development, there is not available an ideal protection method. A convenient method to protect metals is by the use of physical barriers against species such as water, oxygen and hydrogen. In this context, organic coatings have been considered as the most effective protective barriers. In particular, epoxy-based resins are widely applied to protect carbon steel due to easiness of processing and excellent mechanical and chemical resistances (Shin et al., 2010). However, in extended exposures to environment such rigid coatings can fail and once a defect occurs, the corrosive species can reach the metal surface resulting in localized corrosion. Owing to that, frequently the protection with epoxy demands pretreatments or the incorporation of corrosion inhibitors (Radhakrishnan et al., 2009), which may incur in prohibitive extra costs.

A potentially useful method to produce organic coatings with adjustable composition and thickness, on virtually any kind of substrate is the technique known as Plasma Enhanced Chemical Vapor Deposition (PECVD) (Biederman et al., 1992, Yasuda, 1985). Since the characteristics of the films grown by PECVD are strongly influenced by the deposition conditions, it is possible to tailor the properties of the coatings to best fulfill the requirements of a given application. In this work it is discussed the effectiveness of plasma deposited films on the improvement of carbon steel corrosion resistance. The discussions are based on results obtained from a-C:H:Si films deposited onto SAE 1012 samples by glow discharge radiofrequency plasmas generated in atmospheres of oxygen and hexamethyldisiloxane (HMDSO) vapor. With varying the plasma excitation power, the corrosion resistance, determined by electrochemical impedance spectroscopy, increased up to 5 orders of magnitude if compared to the pristine material. Interpretations are provided based on the film thickness, density and roughness.

2. Experimental details

The depositions of the films were performed using the system illustrated in Fig.1. It consists of a cylindrical glass chamber of 7 liters in volume, with two parallel stainless steel electrodes. The lower electrode was used as sample holder while a metallic mesh was used as the upper electrode to facilitate the gas feeding to the inter-electrode region.

Prior to the depositions the samples were cleaned in ultrasonic baths using detergent solution, distillated water and isopropyl alcohol and then the substrates were dried with a hot air gun. Subsequently, they were sputter-cleaned during 600 s. To carry out this process, the chamber was evacuated down to 1 Pa and hydrogen and argon were admitted in the reactor in equal proportion. The total pressure of the gases was 1 Pa and the plasmas were generated by application of radiofrequency power, RF, (13.56 MHz, 150 W) to the sample holder while the upper electrode was grounded. After the plasma cleaning, the depositions were performed during 3600 s in atmosphere of HMDSO and O_2, in equal proportions, by RF application to the sample holder. The total pressure was kept at 20 Pa and the RF power was varied from 50 to 250 W. Figure 2 shows a picture taken from the system during the deposition process.

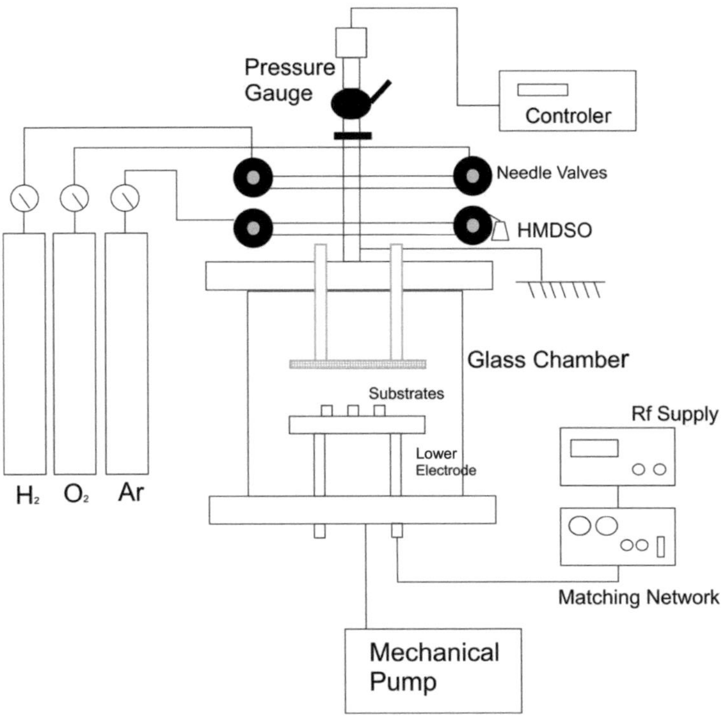

Fig. 1. Experimental apparatus employed in film deposition.

Fig. 2. Picture of the system during the deposition process.

The molecular structure of the films deposited onto polished stainless steel was analyzed by infrared reflectance-absorbance spectroscopy in a JASCO FTIR 410 spectrometer. Spectra were collected with 124 measurements with a resolution of 4 cm^{-1}. The morphology of the carbon steel surface was analyzed by scanning electron microscopy in a Zeiss, EVO MA-15 system. The acceleration voltage was -20 kV and the micrographs were acquired with the secondary electrons detector. The effect of the film deposition on the topography of the carbon steel was investigated by atomic force microscopy in contact mode in a Hysitron Triboindenter. The setpoint was 4.0 μN and RMS roughness was evaluated from the complete image area (50 X 50 μm^2). Surface wettability was assessed from contact angle, θ, measurements using an automated goniometer Ramé-Hart 100. Three drops of deionized water were deposited on different positions of carbon steel plates coated with the plasma deposited film. The results correspond to the average of ten measurements performed with each drop. Film thickness was measured with a profilometer (Veeco, Dektak 150) from a step made onto a glass slide during the deposition, using Kapton adhesive tape as a mask. The deposition rate was calculated by dividing the thickness by deposition time. This instrument also enabled the evaluation of surface roughness of the plasma coated carbon steel plates. All the experiments related to perfilometry were conducted in, at least, three different positions of each sample. Electrochemical impedance spectroscopy (EIS) experiments were performed using a lock-in amplifier (EG&G Instrument, 5210) coupled to a Potentiostat-Galvanostat System (EG&G PAR, 273A), connected to a three-electrode electrochemical cell. Either pristine carbon steel or carbon steel coated with the films was used as the working electrode. A platinum foil was used as counter-electrode and a saturated calomel electrode was used as reference electrode. EIS measurements were obtained at open-circuit potential with the samples in a 0.05 M- NaCl solution with frequency ranging from 10^5 to 10^{-2} Hz, with amplitude of 10 mV. The immersion time ranged from 0 to 10800 s. The oxidation resistance of the surfaces was also evaluated from the etching rate, in oxygen or sulfur hexafluoride plasmas. In both cases, samples prepared onto glass plates were exposed for 1800 s to plasmas generated with excitation power of 50 W and at 1.3 Pa. For each sample, the thickness of the plasma generated step was probed in three different positions. To avoid interferences, for all the investigations conducted here, one exclusive sample was prepared for each different test.

Deviations in the results may occur mainly due to the lack of chemical and physical homogeneity of the employed carbon steel. Such variations are observed to strongly affect the surface properties of a solid providing heterogeneous data. Furthermore, the dependence of the plasma treatment results on the geometrical factors, specifically in this case, the sample radial distance from the center of the electrode, is another relevant aspect that may generate deviations. However, to avoid this problem the samples prepared for each analysis were always placed exactly at the same position with respect to the center of the electrode. Moreover, as all the technical recommendations were strictly observed during the plasma treatments, the results that will be presented here are reproducible, according to previous observations.

3. Results and discussions

The thickness, h, and the deposition rate, R, of the films are presented in Fig.3 as a function of the plasma excitation power, P. There is progressive rise in h with increasing P up to 200 W and a sudden fall afterwards, associated to film delamination. Since the deposition time was maintained constant (3.600 s) in all the experiments, R follows the same trend, reaching its maximum value (~ 300 nm/min) at 200 W. In order to understand such results some plasma phenomena should be considered. First of all, the effect of the excitation power on the mean electron energy and density. The density of electrons with energies high enough to activate plasma species rises with P (Choudhury et al., 2010) if other plasma parameters are kept unchanged. As electrons are the energy carriers in such environments, the probability of bond fragmentation, excitation and ionization through inelastic collisions grows. Plasma activity increases intensifying all the processes taking place as deposition, ablation and ion bombardment.

Ion bombardment during the deposition has important implications for the final properties of the system. As in this work substrates were placed at the driven electrode, under the action of the self bias polarization, positive ions are accelerated through the sheath and implanted in the exposed surfaces. It is well known that self bias voltage scales with the excitation power (Choudhury, et al., 2011) if the gas pressure is kept constant. Thus, the increment in P leads to an increment in the ion accelerating potential which has two effects: it increases the flux of low energy carbon ions towards the substrate surface, affecting the deposition velocity (Morosoff, 1990), and also provides more energy to the newly formed layer what is decisive for the final material properties. Therefore, the slope in the curves of Fig.3 can be attributed to changes in the deposition kinetics due to variations in the intensity of the ion bombardment and in the density/energy of the electrons in the plasma and it is expect that such alterations would change not only deposition rate but also the film characteristics.

The infrared spectra of the films are shown in Fig.4. The wavenumber of the absorptions and their respective assignments are listed in Table 1. The main vibrational modes of the monomer molecule (HMDSO) are identified revealing the incorporation of the organosilicon fragments in the film structure.

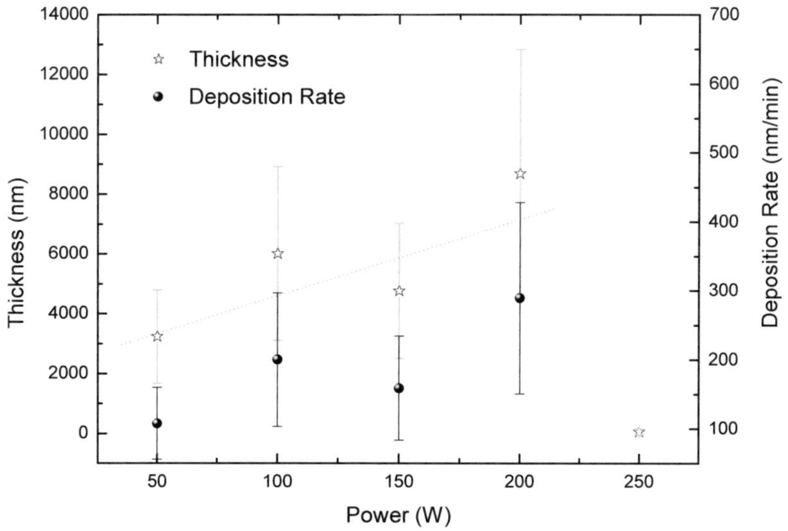

Fig. 3. Thickness and deposition rate of the films as a function of the plasma excitation power.

In the spectrum of the film deposited with the lowest power, 50 W, there are contributions related to C-H stretching (2839, 2902 and 2961 cm^{-1}) and bending (1257, 1354, 1411 and 1440 cm^{-1}) modes. The presence of adsorbed water and carbonyl groups is suggested by the contributions around 1600 and 1727 cm^{-1}, respectively. The retention of silicon in the film is evidenced by absorptions centered at 1020 (ν Si-O) and 1130 (ν Si-O) cm^{-1}. In the low wavenumber region, other contributions also ascribed to silicon-containing groups, appear around 700 (δ Si-H$_n$), 755 (ν C-H in $Si(CH_3)_3$), 782 (ν Si-C), 806 (ν Si-O, ν Si-C and δ CH_3), 836 (ν Si-C) and 851 (δ CH_3 $Si(CH_3)_3$) cm^{-1}.

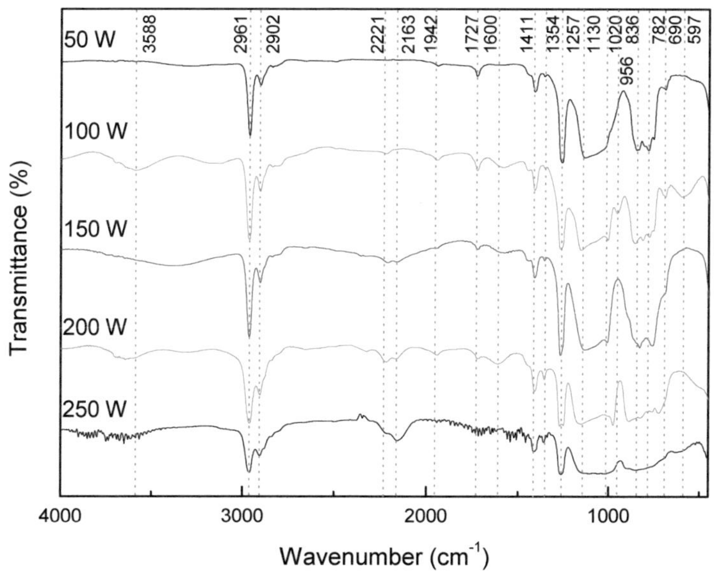

Fig. 4. Infrared spectra of films deposited in plasmas of different powers.

All the above mentioned absorptions are evident in the spectra, but modifications are detected with increasing P. In the spectrum of the sample prepared with 100 W, there is emergence of bands laying around 3588 and 956 cm^{-1}, respectively related to the stretching and bending modes of O-H vibrations in silanol groups. Once O-H is not originally present in the structure of the organic molecule, the observation of this group indicates multiple step reactions in the plasma phase. In the spectra of the films prepared with 150 and 200 W, these vibrations are no longer detected, but they retain large bands centered at 3370 (150 W) and 3636 cm^{-1} (200 W). While the first is characteristic of intermolecular hydrogen bonded O-H, the second is ascribed to O-H in primary alcohols ($R-CH_2-OH$) (Scheinmann, 1970), reveling alteration in the hydroxyl form of incorporation. The depletion of silanol groups can also be associated to an increment in the crosslinking degree (Fracassi et al., 2003).

Wavenumber (cm^{-1})	Assignments	Groups	
690, 700	$Si-H_n$		(Guruvenket et al., 2010)
755 782	δ $C-H_3$ ν Si-C	$Si-(CH_3)_3$ $Si-(CH_3)_2$	(Rao et al., 2010) (Guruvenket et al., 2010)
806	ν Si-O ν Si-C, δ CH_3	Si-O-Si $Si(CH_3)_x$	(Choudhury et al., 2010) (Gengenbach et al., 1999, Scheinmann, 1970)
836 851	ν Si-C δ $C-H_3$	$Si-(CH_3)_x$ $Si-(CH_3)_3$	(Rao et al. 2010) (Gengenbach et al., 1999)
956	δ O-H	Si-OH	((Ricci et al, 2011, Choudhury et al., 2010)
1020, 1130	ν Si-O	Si-O-Si	(Gengenbach et al., 1999, Fracassi et al., 2003)
1257	δ $C-H_3$	$Si-(CH_3)_x$	(Gengenbach et al., 1999, Choudhury et al., 2010)
1354	δ $C-H_2$	$Si-CH_2-Si$	(Ul et al., 2000)
1411, 1440	δ C-H	$Si-(CH_3)_x$	(Rao et al., 2010, Ul et al., 2000)
1600	O-H	Free Water	(Rao et al., 2010)
1727	ν C=O	CH_2O	(Ricci et al., 2011)
2163, 2221	ν Si-H		(Ul et al, 2000)
2902	ν C-H	CH_2	(Gengenbach et al., 1999)
2961, 2839	ν C-H	CH_3 and CH_2	(Gengenbach et al., 1999, Scheinmann, 1970)
3588	ν O-H	Si-OH	(Ricci et al., 2011, Choudhury et al., 2010)

Table 1. Assignments of the bands in the infrared spectra of the films and their related groups. Symbols ν and δ represent stretch and deformation vibrations, respectively.

Still in the spectrum of the film deposited with 100 W, there appears bands at 2163 and 2221 cm^{-1}, ascribed to Si-H stretching in good agreement with the presence of the band related to Si-H bending mode (~ 700 cm^{-1}). The intensity of all Si-H contributions increases with P, indicating progressive growth of the dissociation and recombination processes in the plasma phase. Figure 5 presents the trends in the relative density of Si-H bonds as a function of P, evaluated from the model proposed by (Lanford and Rand, 1978). In fact the Si-H proportion tends to increase with P, but the growing rate was observed to be higher for the films deposited with 150 and 250 W.

The absorption related to Si-O vibrations, at 1020 cm^{-1}, shifts to lower wavenumbers with increasing P indicating film densification what is also consistent with higher proportions of Si-H groups (Rao et al., 2010) in the structure. This effect is clearly evidenced in Fig.6 which highlights the bands in the low wavenumber region. On the other hand, the intensification of this contribution (up to 200 W) reflects the increment of Si-O-Si fraction and therefore, of the crosslinking degree (Gengenbach et al., 1999). Thus, both changes observed in the band around 1020 cm^{-1} are consistent with a silicon enrichment with increasing P. The noticeable widening and loss of resolution of all the bands arising in the low wave number region (< 1000 cm^{-1}) suggest changes (disorder) in the neighborhood of the chemical bond due to a higher diversification of fragments incorporated in the structure.

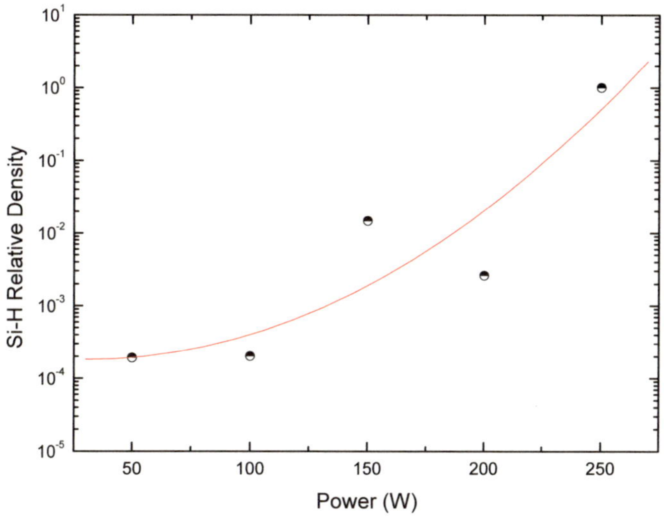

Fig. 5. Relative density of Si-H bands in the film as a function of the plasma excitation power.

An evolution is also observed in the bands related to C-H deformation in CH_3 (1411 cm^{-1}) and CH_2 (1354 cm^{-1}) groups. Whereas the first keeps roughly unchanged, the second increases with P. The growth of the bands related to CH_2 species is an evidence that fragmentation of $Si(CH_3)_3$ groups is an important route of the polymerization process, differently of the results found in the work of Gengenbach (Gengenbach et al., 1999). Indeed, such a mechanism represents one of the major fragmentation process in HMDSO plasma processes. The overall intensity of the band around 1260 cm^{-1} (CH_3 in $Si(CH_3)_x$) is also changed in the spectra of Fig.6: it is observed to decrease with P despite the increment of the film thickness. This result indicates continuous depletion of $Si(CH_3)_x$ groups, probably by methyl or hydrogen abstraction, generating dangling bonds were Si-H groups can be formed. According to Rao (Rao, 2010), the reduction in the intensity of this band is associated to the increase in the structure crosslinking.

Even though thickness results indicated complete film detachmentfrom the sample prepared at 250 W, infrared spectrum detected the presence of the layer. After careful visual inspections it was observed that film detached of some substrates and was preserved in others. The aging time upon atmospheric conditions also influenced the adhesion of this sample to the substrates.

Figure 7 shows the contact angle, θ, of the films as a function of the plasma excitation power. All the surfaces are hydrophobic presenting contact angles higher than 90°. The film which presented the lowest θ value was the prepared with 100 W. It is interesting to point out that this was the only sample in which silanol groups were detected in the infrared spectra (956 and 3588 cm^{-1}). The presence of such species is normally associated to points where hydrolization can occur upon exposition to atmosphere or to water solutions, providing an early deterioration of the barrier properties. For P > 100 W, there is no significant variation of θ which keeps around 104°.

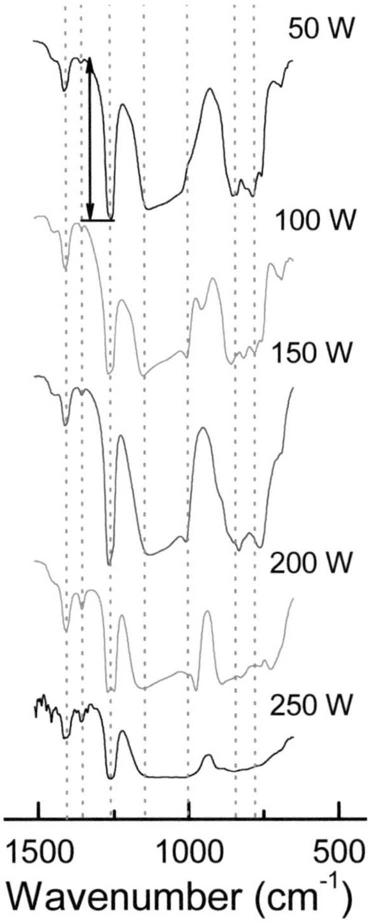

Fig. 6. Infrared spectra of the films prepared in plasmas of different power in the low wavenumber region.

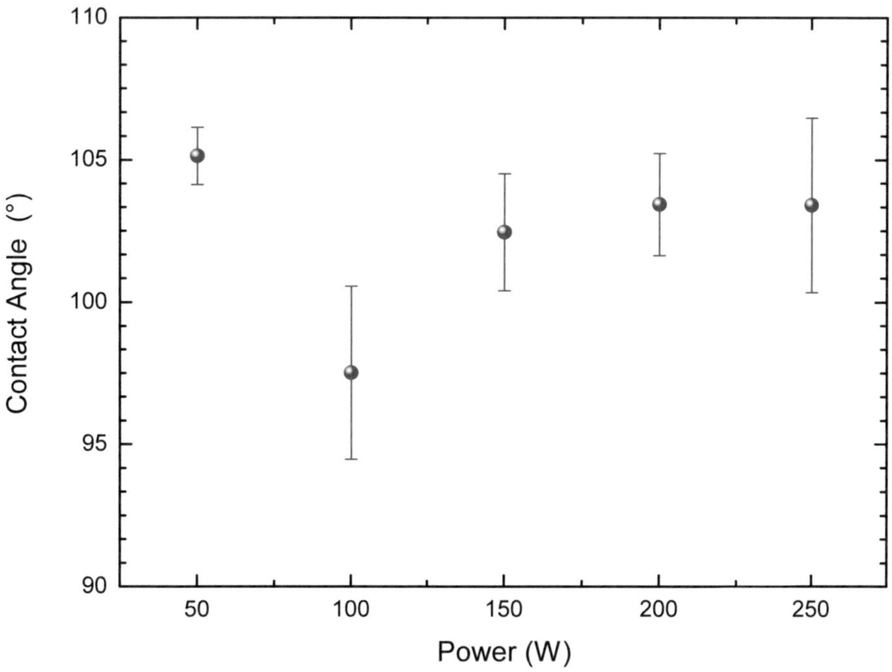

Fig. 7. Contact angle of the films as a function of the plasma excitation power.

It is interesting to observe the extreme difference in the water wettability as one consider silicon oxide and plasma deposited HMDSO films. In the first, the high concentration of polar Si-O groups attains high surface energy to the material, stimulating the incorporation of O-H atmospheric groups. These groups favor the connection of atmospheric water molecules through hydrogen bonds, affecting the way water spreads onto the surface. Even though films prepared from HMDSO plasmas normally present high proportions of Si-O groups, they are shielded by the non polar methyl groups, reducing the electrostatic force over water molecules. Therefore, as high the proportion of trimethylsylil in the surface, higher the contact angle, what is in good accordance with the results of Fig.7: the film prepared with the lowest power associated the highest $Si(CH_3)_x$ proportion to the highest θ value.

Another important aspect as one considers water wettability is surface topography. Changes in roughness alter the contact area of the droplet varying the intensity of the electrostatic forces. Figure 8 shows the roughness of the as-received (dotted line) and plasma deposited carbon steel as a function of P. In a general way, film deposition reduces the substrate roughness for any deposition condition employed here. Furthermore, the roughness upshifts just for the samples prepared at 150 and 250 W of power.

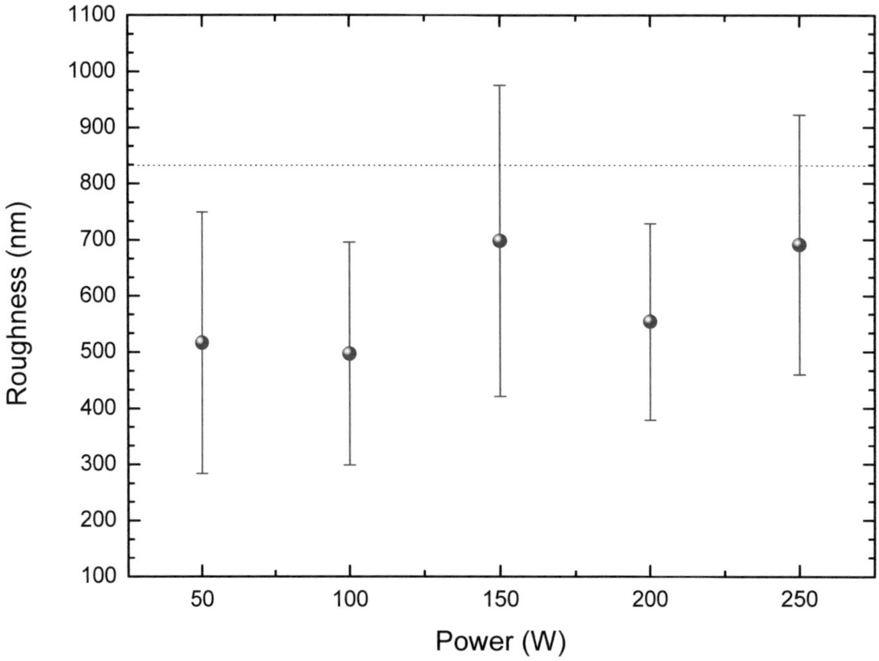

Fig. 8. Roughness of the films as a function of the plasma excitation power. Dotted line represents the roughness of the bare carbon steel.

To evaluate the effect of the roughness on the water wettability, the graph of Fig.9 was built. It is well accepted in the literature that for hydrophobic materials the increment in surface roughness intensifies the repulsive electrostatic forces due to the increase in the contact area (Quéré, 2005). However, films prepared here are hydrophobic by the absence of electrostatic forces and not by intense repulsive interactions. Therefore, it is expected that roughness increment could expose Si-O polar groups by structural rearrangements, increasing their electrostatic interactions with polar water molecules. In this case, roughness increment would tend to reduce θ. Through the results presented in Fig.9, it is indeed observed a slight fall tendency in θ with increasing roughness above 500 nm, suggesting interference of the relief on the contact angle. But the topographical changes do not explain the steep θ slope for the first two points in the graph, since roughness is practically constant for them (497 ± 198 and 516 ± 233 nm). In fact, the lowest θ value presented by the smoothest sample (100 W) coincides with the appearance of silanol groups in the structure. Therefore, plasma power affects wettability of the films due to the chemical and topographical alterations, being the first one the most prominent in the definition of the trends.

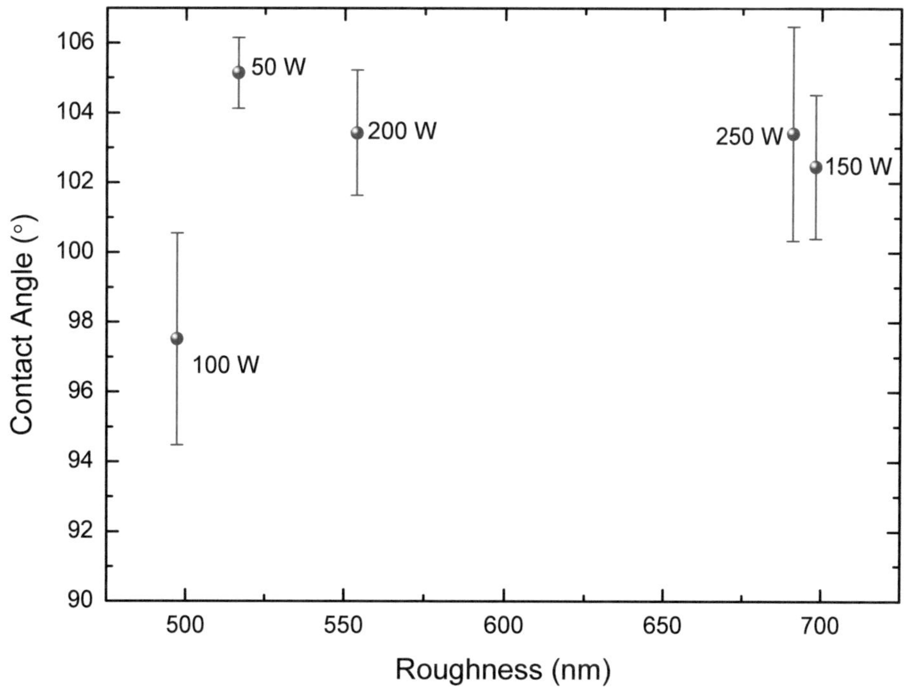

Fig. 9. Contact angle as a function of the roughness of the films.

The corrosion of Si-containing films upon fluorine- and oxygen–rich environments is a matter of interest in microelectronic (Chen et al., 2011, Stillahn et al., 2011) and aerospace (Huang et al., 2011) applications, respectively. To evaluate the chemical response of the films to such media, samples were exposed to reactive plasmas generated from SF_6 or O_2. The etching rates derived from these experiments are presented in Fig. 10 as a function of the film deposition power. The removal of the layer material in SF_6 plasmas is slower for the films prepared with 50 and 100 W and rises around 30% with increasing P to 150 and 200 W. Since the layers prepared with the highest power levels presented progressive loss of C-H groups and then silicon enrichment, structures with elevated proportions of Si-H and Si-O groups were formed. The affinity towards fluorine grows, consistently with the etching rate elevation. This interpretation is supported by the fact that etching rates similar to the reported to silicon wafers in SF_6 plasmas (~ 90 nm/min) (Tian et al., 2000) were attained to the high power prepared samples. As materials with elevated etching rate and selectivity are desired for designing high aspect ratio microelectrical-mechanical systems, the films prepared with the highest power levels are potential candidates for this area.

It is interesting to note, however, that when the samples were exposed to oxygen plasmas, no material removal was detected by perfilometry indicating the formation of a structure highly resistant to oxygen attack, as required in several aerospace applications. Even though C is present in huge proportions, it is protected from oxygen attack by formation of a $Si\text{-}O_2$ barrier on the film surface upon oxygen plasma (Bruce et al., 2010). From the moment this protection is created, further material removal is hampered, explaining the undetectable etching rates developed in the films studied here.

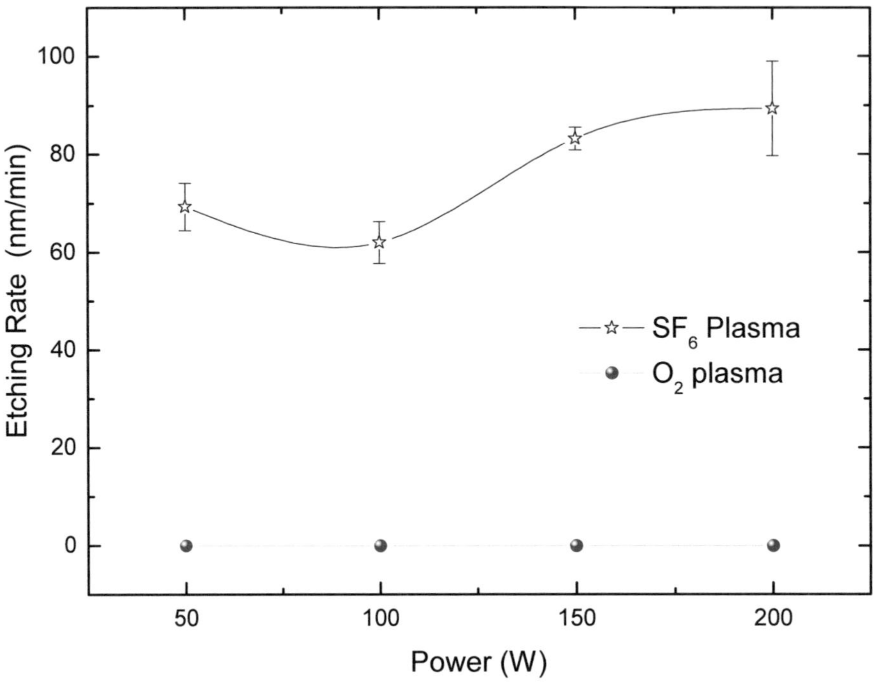

Fig. 10. Etching rate of the films in reactive SF_6 and O_2 plasmas as a function of the film deposition power.

The resistance of the carbon steel to corrosion in saline solution was evaluated by electrochemical impedance spectroscopy. The results of the phase angle are presented in Fig.11 as a function of frequency, ν, for the bare and film-containing carbon steel. Measurements were performed varying the immersion time, t, from 0 to 180 minutes.

Considering the immersion time t = 0 min, the phase angle of the non-treated substrate increases with frequency reaching a maximum around 10^1 Hz and falling afterwards. This one concavity shape reflects the changes in the phase angle provoked by substrate-electrolyte interactions. After film deposition, the low frequency maximum is still detected but a more intense component arises in the high frequency region, affecting the overall shape of the curves. The second maximum is caused by electrolyte- film interactions and the presence of the two components suggests simultaneous reactions on the film and substrate surfaces. The enhancement in the high frequency component, accompanied by the diminution in the low frequency one, reveal a better performance of the system under the corrosive medium. For the sample prepared with 250 W no improvement was detected with respect to the bare steel since the film partially detached of the substrate. Comparing the phase angles at 10^5 Hz in the different graphs, it is noticed an evolution with increasing P: for the bare substrate it is lower than 10° and increases continuously, reaching 86° for the sample prepared in plasma of 200 W. Layers with phase angles close to 90° present a capacitive-like behavior, and act as good corrosion barriers. For the sample prepared with 150 W, the phase angle at 10^5 Hz is high but not stable with increasing t: a continuous fall is observed, reaching the same plateau detected for the bare substrate. Indeed, the shape of

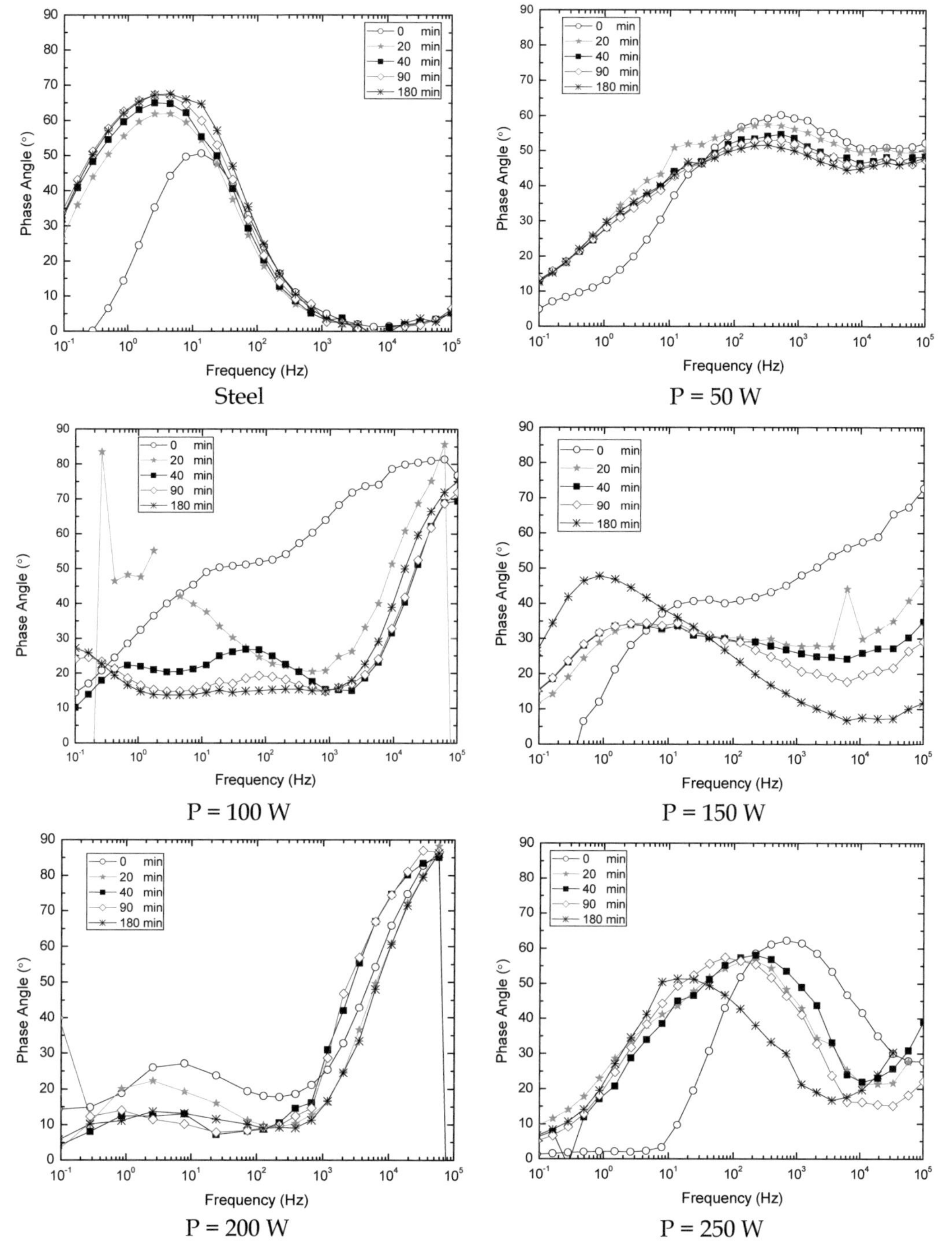

Fig. 11. Phase angle as a function of the frequency taken at different immersion times for the samples prepared on carbon steel in plasmas of different powers. Phase angle for the bare steel substrate is also presented.

this curve evolutes to that of the bare substrate if the test is conducted for 180 minutes. Although the film deposited with 100 W presented good results for zero immersion time, they were not preserved with increasing t. The treatments which presented the most stable results were those performed with 50 and 200 W, being the last one, the most effective for the corrosion protection of the carbon steel.

Figure 12 shows the impedance modulus as a function of frequency for the bare- and film-containing carbon steel. In a general way, |Z| decreases with increasing ν, achieving the lowest impedance at the high frequency extreme. Independently of the deposition condition employed, film application upshifts the impedance curves, especially in the low frequency region. Considering the different immersion times, the samples exposed to 50 and 200 W plasmas presented high stability even after the longest test. On the other hand, a downshift is observed on the curves of the samples prepared in plasmas of 100, 150 and 250 W, indicating a deterioration of the protective properties upon the corrosion experiments. The lack of a specific trend with t in the curves of the sample deposited at 250 W may be a consequence of the non uniformity of the remaining layer left onto the substrate.

Using the method proposed in (Mansfeld, 1981), the total resistance R_t, was derived from the impedance modulus curves at the lowest frequency extreme (10^{-1} Hz) and the results are presented in Fig.13 as a function of the immersion time. Comparing the first point of each curve (t = 0 min) it is readily observed that the film deposited at 250 W provided no further resistance to the carbon steel. On the other hand, the film prepared with P = 150 W enhanced R_t by 2 orders of magnitude. Still better results were encountered for the samples exposed to plasmas of 50, 100 and 200 W. In these cases, R_t reached 1.5 X 10^6, 2.5 X 10^7 and 3.1 X 10^6 Ω, respectively, values substantially higher than the obtained for the bare substrate (5.3 X 10^2 Ω). Nevertheless, these high impedance values decreased with increasing immersion time to 20 minutes, and kept practically unchanged afterwards. This constancy at elevated values is an important issue as one considers the preservation of the barrier properties under aggressive environments. Under the most aggressive condition employed here (180 min), R_t was also observed to vary with P: it increases from 1.5 X 10^5 to 5.4 X 10^5 Ω as P is changed from 50 to 200 W, that is an almost fourfold increase for this P range.

In order to interpret the results of Fig.13 it should be taken into account the effect of the film thickness on the corrosion protection, as depicted in Fig.14. The growth in R_t scales well with film thickness, except for the sample prepared in 150 W plasmas. In this case, another factor, besides thickness, seems to have a greater importance for the film performance.

Composition is also a relevant factor for corrosion resistance. In plasma deposited HMDSO films (Ul et al., 2002), the density is observed to increase with the proportion of Si-H bonds due to Si enrichment. From the results presented in Fig. 5, the Si-H proportion augments in the film with increasing P. According to this analysis, the densest layers were prepared in 150 and 250 W plasmas while the lightest ones were obtained in 50 and 100 W plasmas. Interestingly, in these extremes Si-H contents are separated by four orders of magnitude.

To determine the effect of the silicon content on R_t, the graph of Fig. 15 was built. Indeed, R_t tends to decrease with increasing Si-H density, except for the sample prepared at 200 W which associated intermediary Si proportions to high thickness values, factor that was determinant for the results found here.

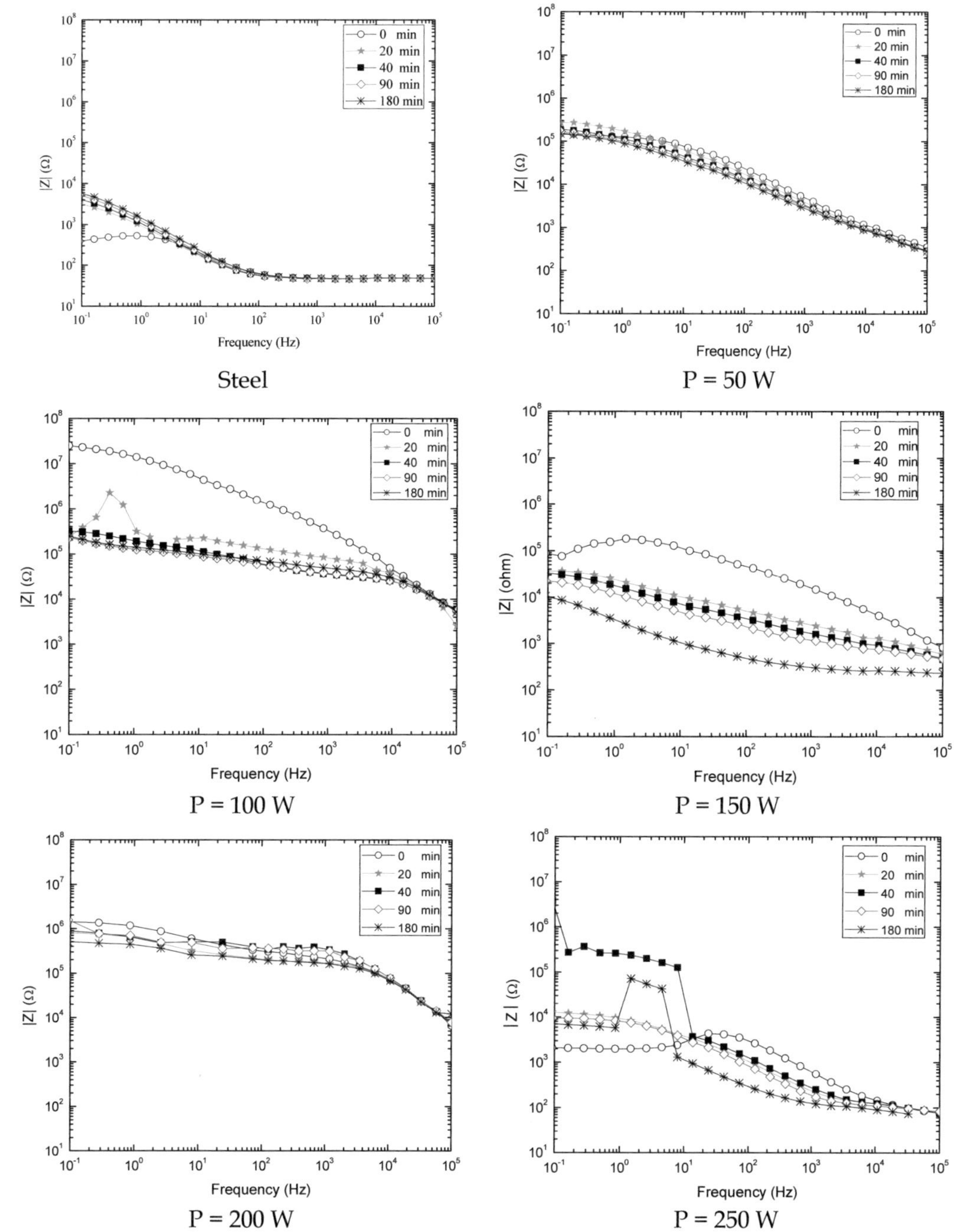

Fig. 12. Impedance modulus as a function of frequency taken at different immersion times for the samples prepared on carbon steel in plasmas excited with different powers. Impedance modulus for the bare steel substrate is also presented.

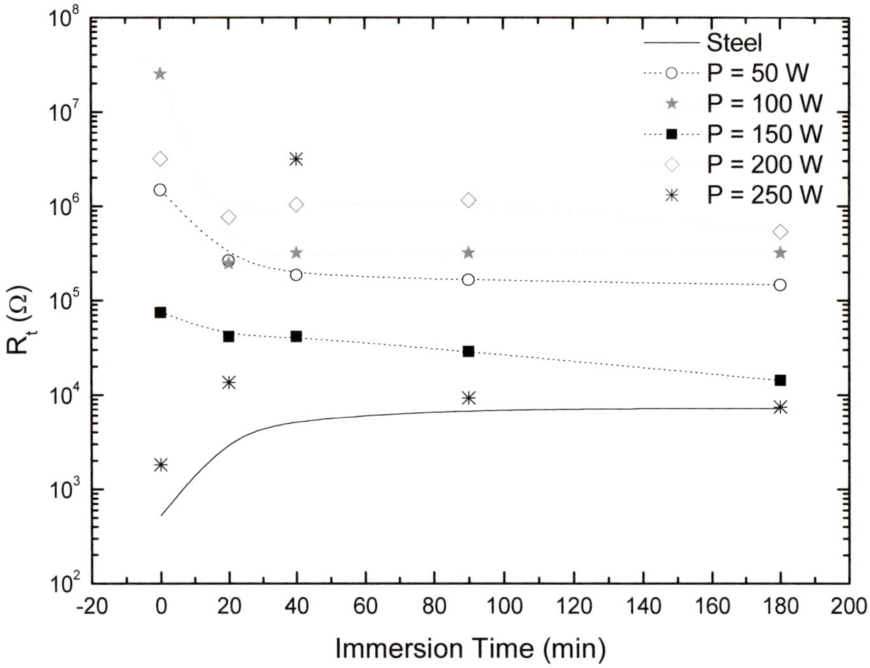

Fig. 13. Total resistance as a function of the immersion time for the samples exposed to plasmas of different powers and for the bare substrate.

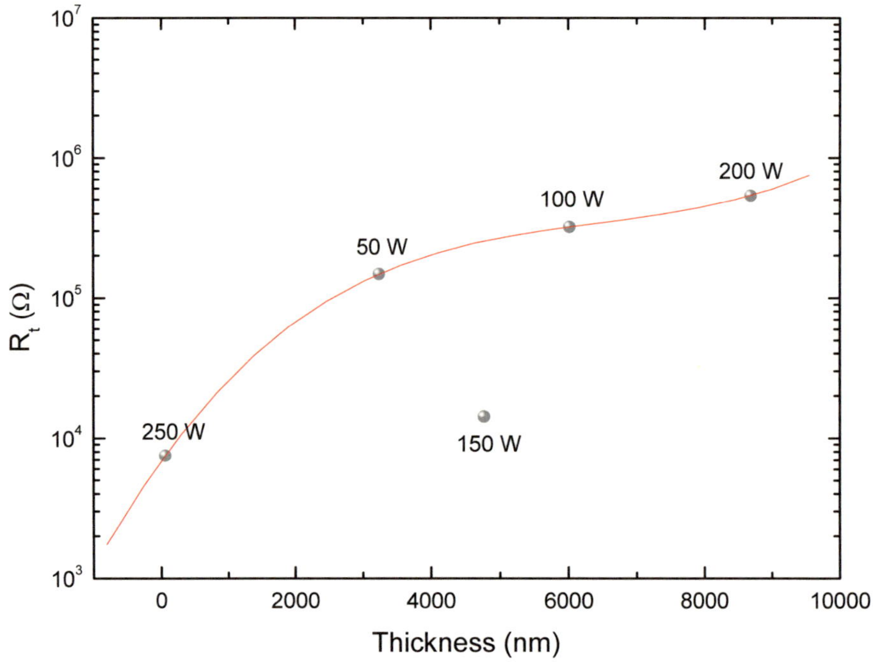

Fig. 14. Total resistance at t = 180 minutes as a function of the film thickness.

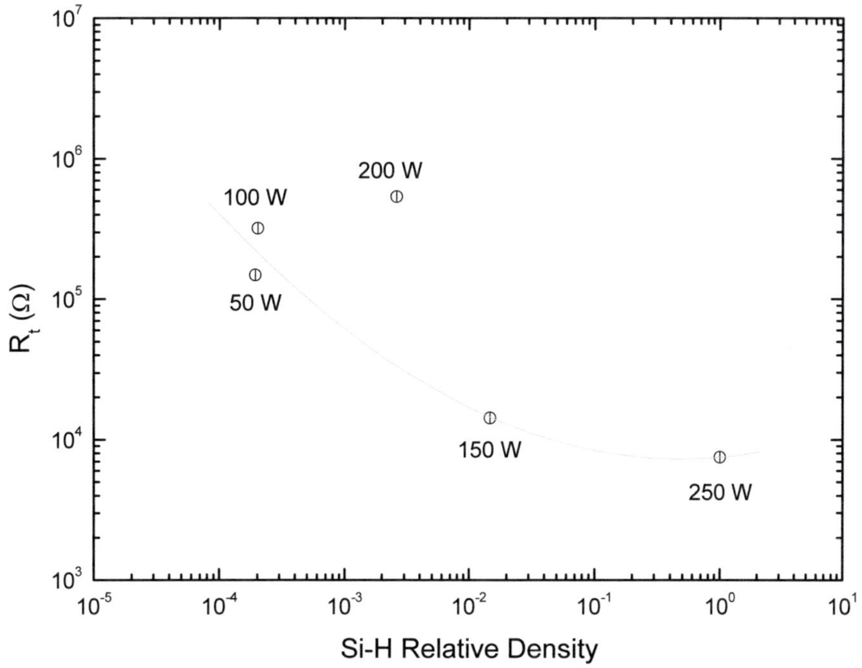

Fig. 15. Total resistance at t = 180 minutes as a function of the Si-H relative density in the films.

Therefore, the poor corrosion performance of the films prepared with 150 and 250 W is ascribed to the combination of the low thickness and high Si contents of these samples. Besides, in the layer deposited with 250 W, cracks, pinholes or other kind of defects may originate from the partial film detachment, producing points were the electrolyte can easily diffuse through the layer reaching the substrate. Films deposited with 50 and 100 W attained low densities of Si-H groups and intermediary thicknesses. Finally, the layer which combined high thickness and moderate concentrations of Si-H species (200 W) presented the highest total resistance and physical stability upon the saline solution.

Scanning electron micrographs of the surfaces, taken prior and after the corrosion experiments, are presented in Fig.16. The bare substrate presents a series of defects throughout the surface, originated of the manufacturing process. After the corrosion tests, such defects are still evident together with residues of the electrolyte-steel reactions. In a general way, film deposition hides the majority of the substrate defects, explaining the roughness fall detected by perfilometry (Fig.8). Just shallow holes and scratches remain on the surfaces indicating incomplete coverage of the deepest imperfections. The structure is formed by the agglomeration of particulates giving rise to a granular film. The dimension of the particulates is observed to increase with P up to 100 W and to decrease afterwards. However, this dimensional alteration was not observed to influence the roughness results of Fig. 8.

	Prior Corrosion	After Corrosion
Steel	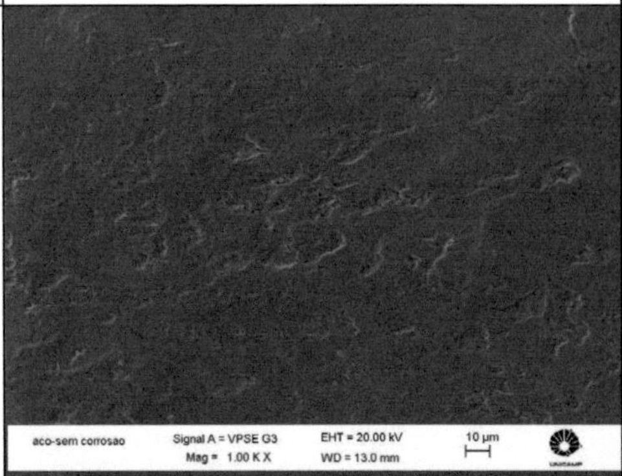	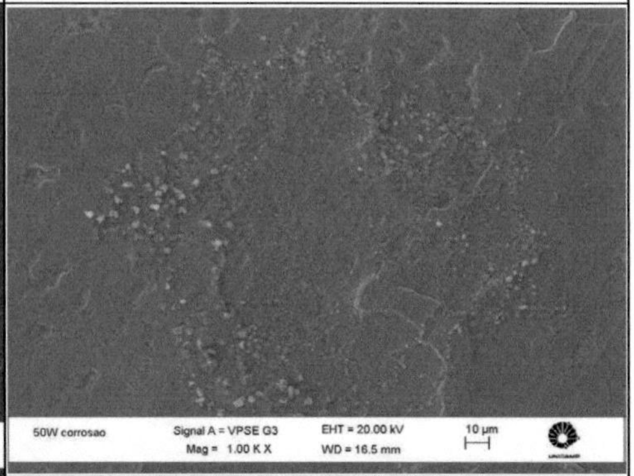
50 W	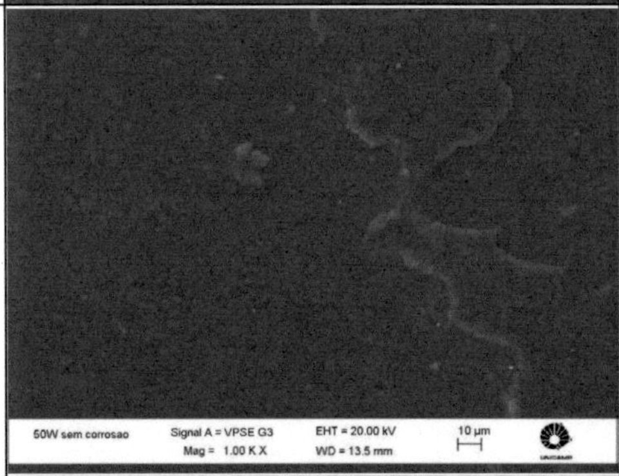	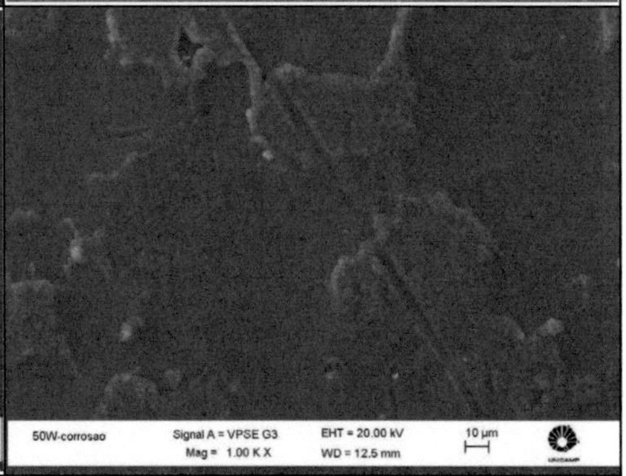
100 W	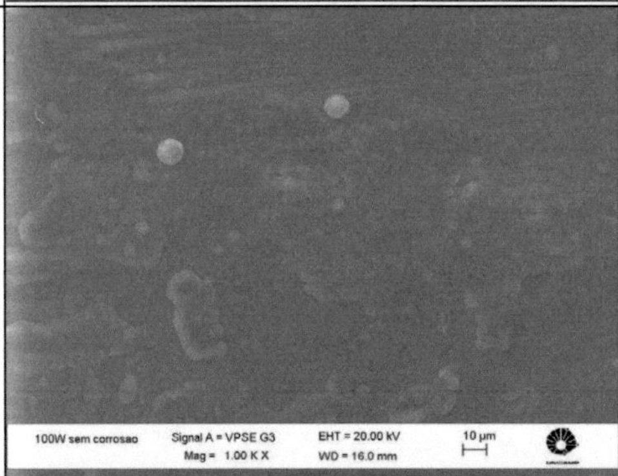	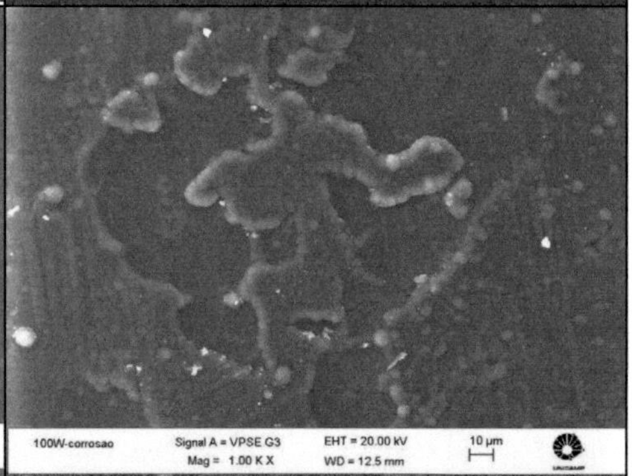
150 W	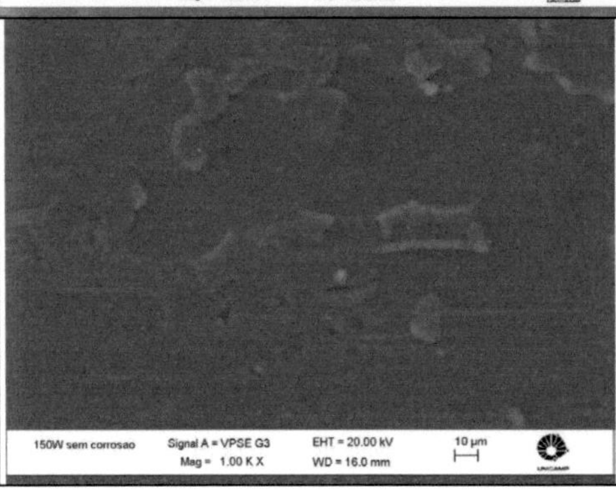	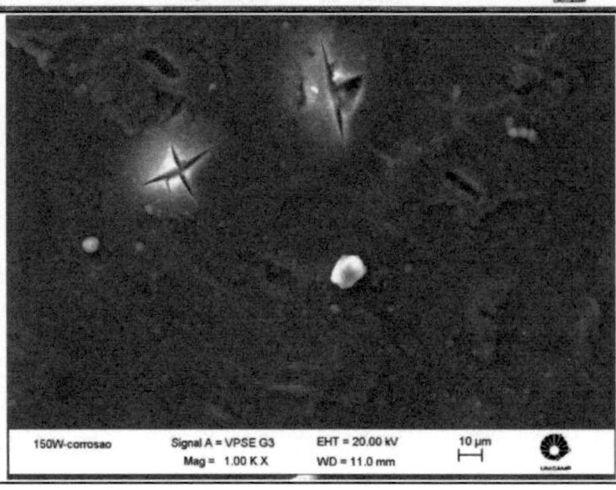

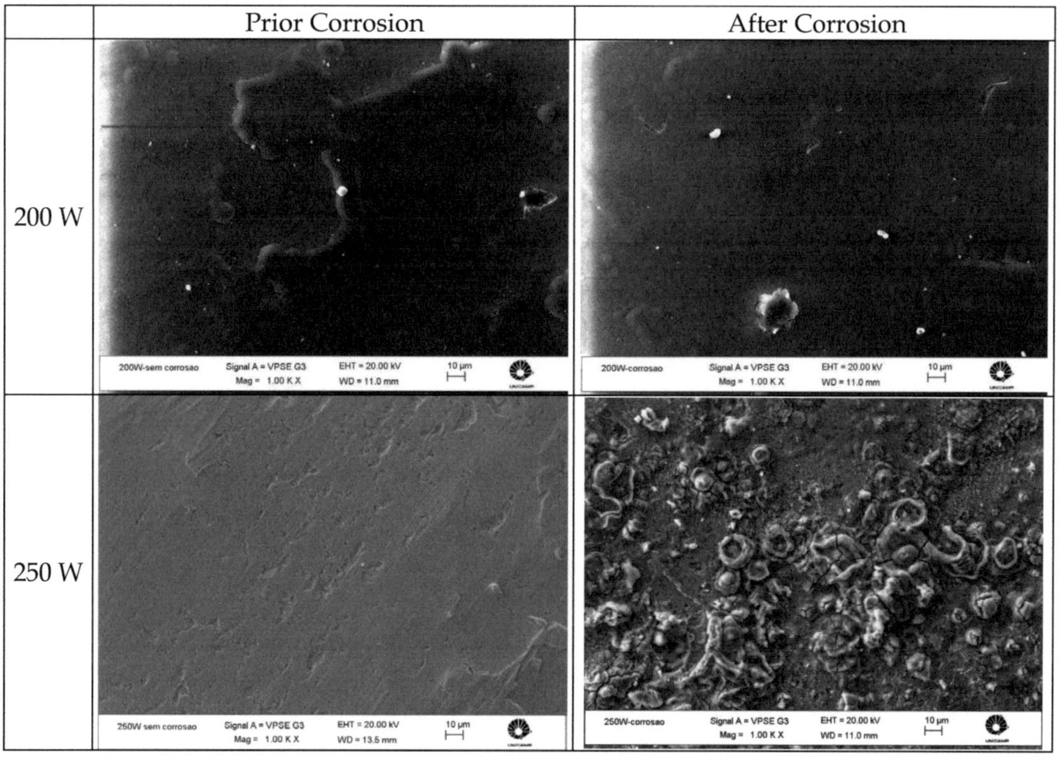

Fig. 16. Micrographs of the samples prepared onto carbon steel in plasmas excited with different powers prior and after corrosion experiments. Micrographs of the bare substrate is also presented for comparison.

Even though structural defects are evident in the films, the overall surface aspect is not changed after the corrosion tests suggesting the presence of the film in their deepest part. Just the sample containing the film deposited with 150 W presented some star-like cracks that may be responsible for its early fail in the EIS tests. In this case, the high density may generate fragile points in the structure that breaks upon solution absorption. The sample deposited with 250 W presented the same characteristics of the bare steel, indicating complete film detachment from the analyzed region. This also explains the similar appearance of the EIS curves for the substrate and 250 W treated sample in Fig.13. For this sample, the post corrosion surface image highlights the film residues and corrosion byproducts precipitated on the surface.

To further evaluate the effect of the power on the relief and morphology of the samples, atomic force microscopy images were acquired and are presented in Fig.17. Line profiles of specific regions are also depicted. Through the information of the image and of the graph line it is possible to verify the kind and dimension of the irregularities in the as-received substrate. After film deposition, cracks, holes and depressions are still detected suggesting that film application was not enough to hide the substrate imperfections. An uniform film containing particles sparsely immersed on it was verified for the sample exposed to 50 W plasmas. The morphology is strongly altered with increasing power to 100 and 150 W.

<table>
<tr><th></th><th>(a)</th><th>(b)</th></tr>
<tr><td>Steel</td><td>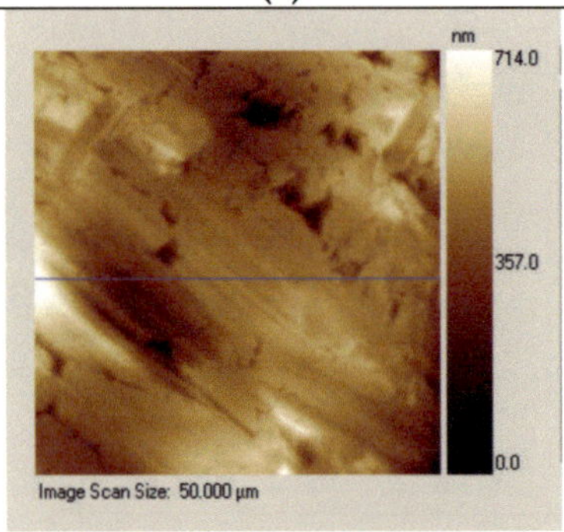</td><td>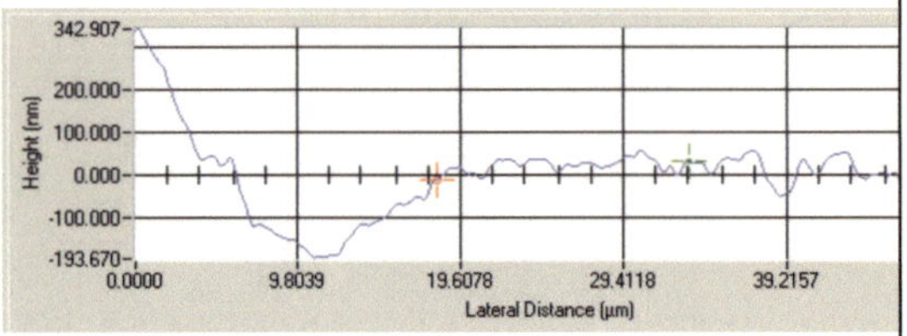</td></tr>
<tr><td>50 W</td><td>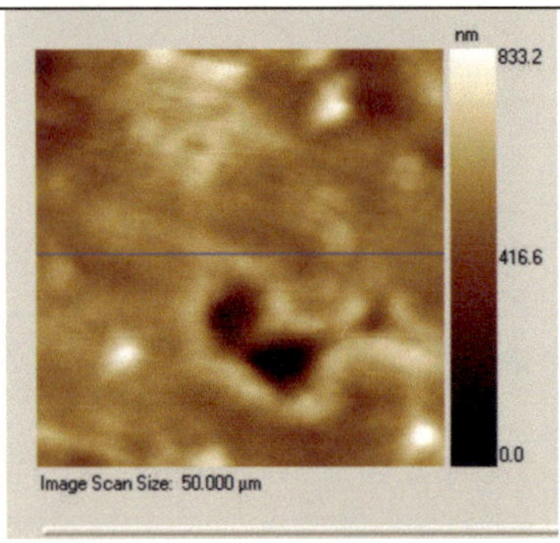</td><td>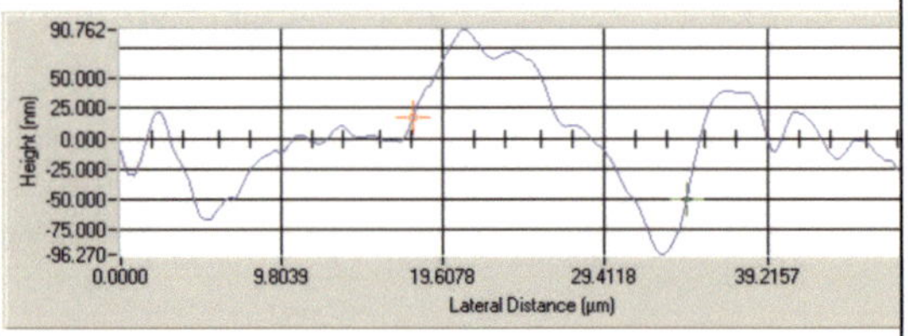</td></tr>
<tr><td>100 W</td><td>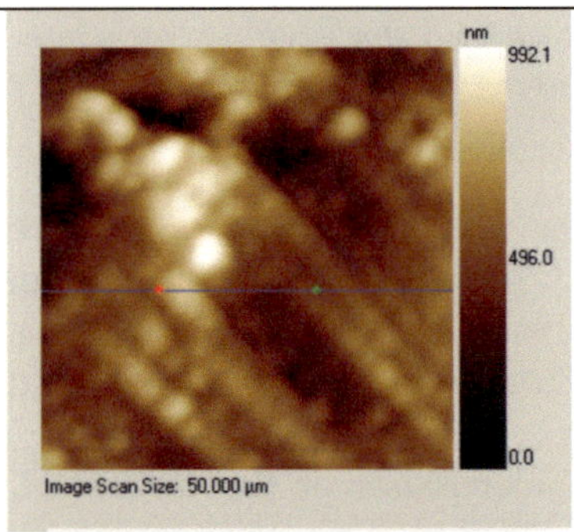</td><td>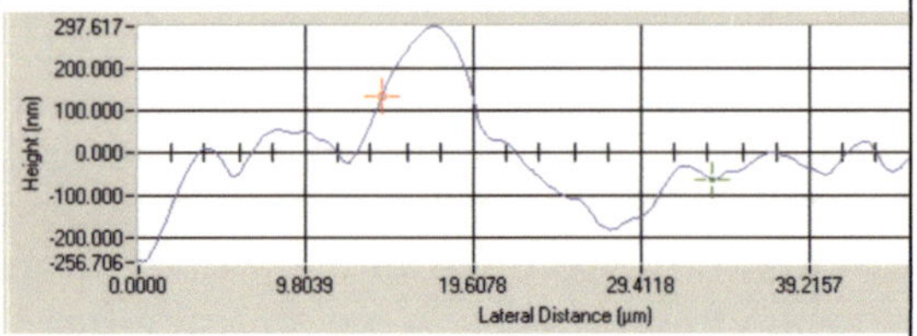</td></tr>
<tr><td>150 W</td><td>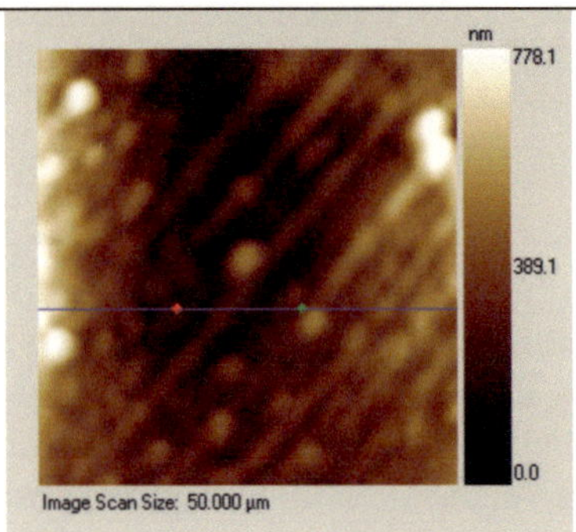</td><td>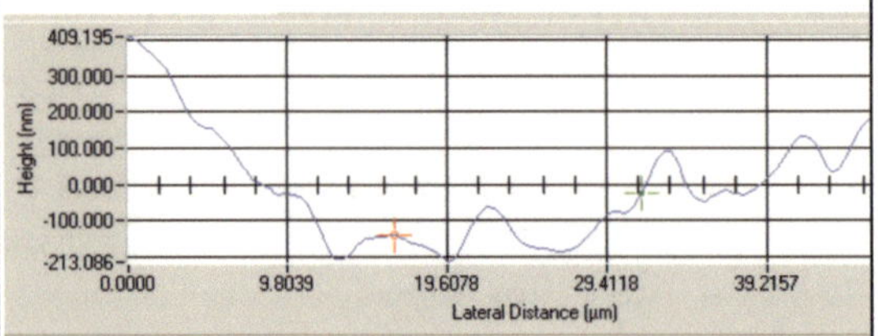</td></tr>
</table>

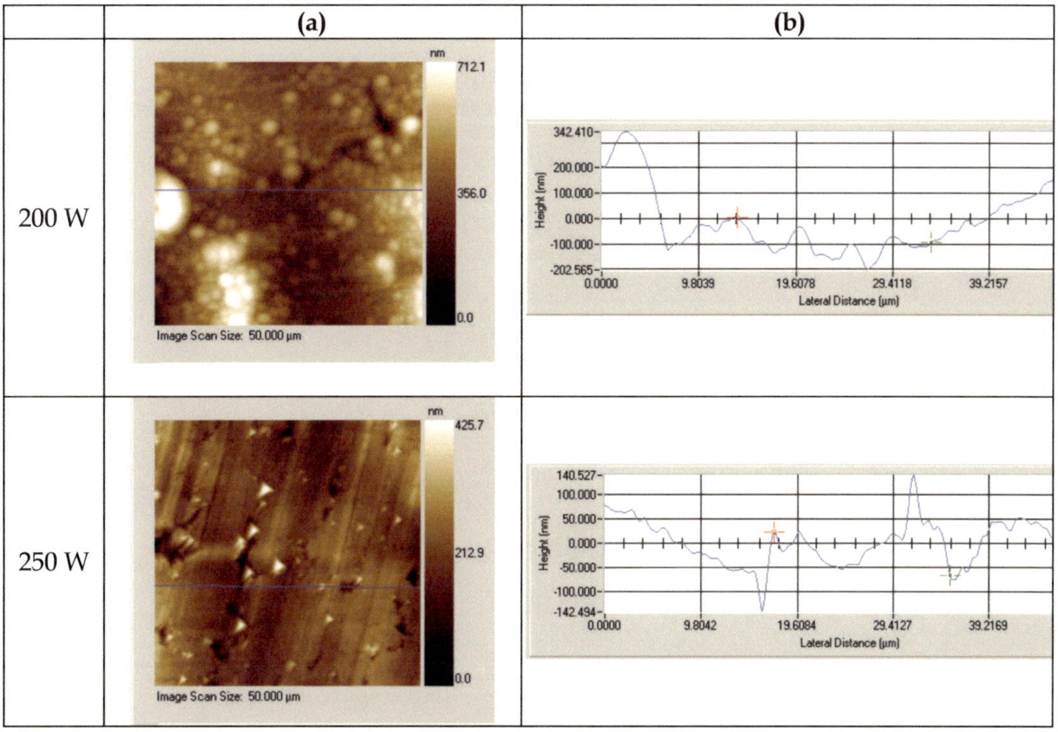

Fig. 17. (a) Surface topographical images of the carbon steel exposed to plasmas of different powers. The image of the non coated substrate is also presented. (b) Surface profiles taken in specific points of each sample, indicated by the blue line in the respective image.

Particles of different sizes are arranged in a string-like granular structure. As power is enhanced to 200 W, the regularity of this pattern is lost and the compact organization of the particles suggests that there are connections amongst them. In this case, the stripes are detected only in the depressions and the defect evident in the image is a result of the incomplete covering of the substrate imperfections. Certainly, film is present in its bottom part, preventing early corrosion fail in this sample. The lack of organization may be a consequence of the higher fragmentation degree induced in the plasma phase with increasing P. Finally, the results obtained in the sample exposed to 250 W plasmas confirm that film was completely detached from the surface, remaining just small fragments connected to it.

Roughness was also evaluated from the 50 X 50 μm^2 atomic force microscopy images and the results are plotted as a function of the plasma excitation power in Fig.18. The interval delineated by the dotted lines in the graph represents the range of roughness values for the bare substrate. Film deposition tended to increase the roughness of the carbon steel for any plasma excitation condition. Considering the scan length adopted in this case, the effect of the substrate imperfections on the RMS roughness is reduced as compared to the obtained in the results of Fig.8, determined by perfilometry. Even thought surface morphology is strongly affected by changing the plasma excitation power it has no evident influence on roughness which keeps roughly constant around 133 nm.

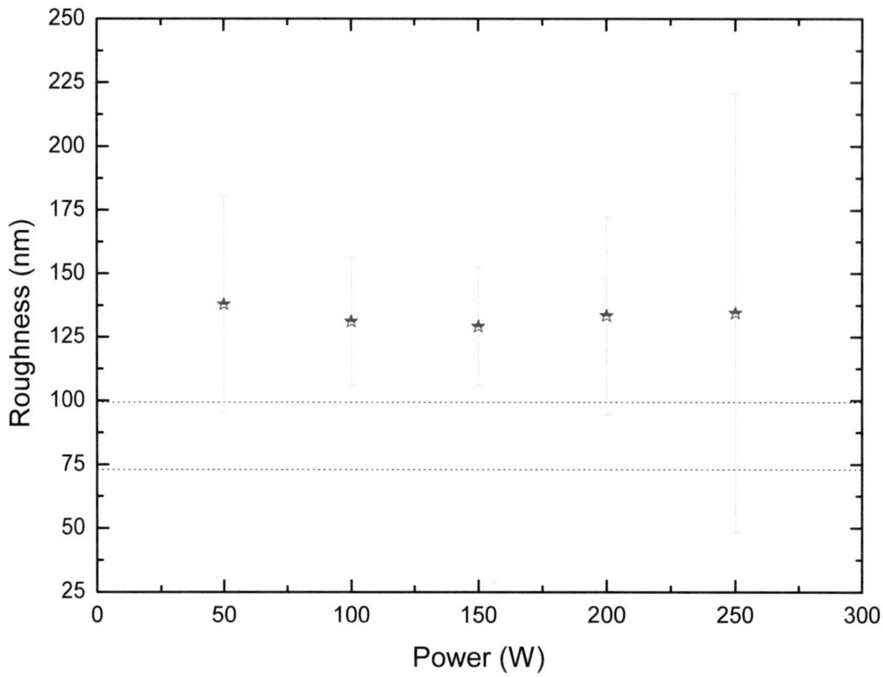

Fig. 18. Roughness of the films as a function of the plasma power. Results obtained from 50 X 50 μm² atomic force microscopy images. Region delimited by dotted lines represents the range of roughness for the bare carbon steel.

The behavior of the total resistance with roughness is presented in the graphs of Figs.19 (a) and (b). In Fig.19 (a) R_t is plotted versus roughness derived from AFM images. No clear correlation is observed amongst the parameters in this picture. However, as R_t is presented as a function of the roughness evaluated from the perfilometry profiles, Fig.19 (b), the results are separated into two groups: the first relative to the smoothest samples which presented high R_t values and the other correspondent to the roughest samples which revealed low R_t values. These results indicate that, besides thickness and silicon proportion, roughness also plays an important role on the performance of system in corrosive media. Therefore, if polished substrates were employed, still higher R_t values would have been expected.

Therefore, the improvement by more than 6.000 times in the corrosion resistance of the carbon steel after the longest corrosion experiments is attributed to the association of high thickness values, moderate silicon contents and low roughness of the film prepared in 200 W plasmas. The integrity of this coating even after the corrosion test is another important aspect as one considers practical applications.

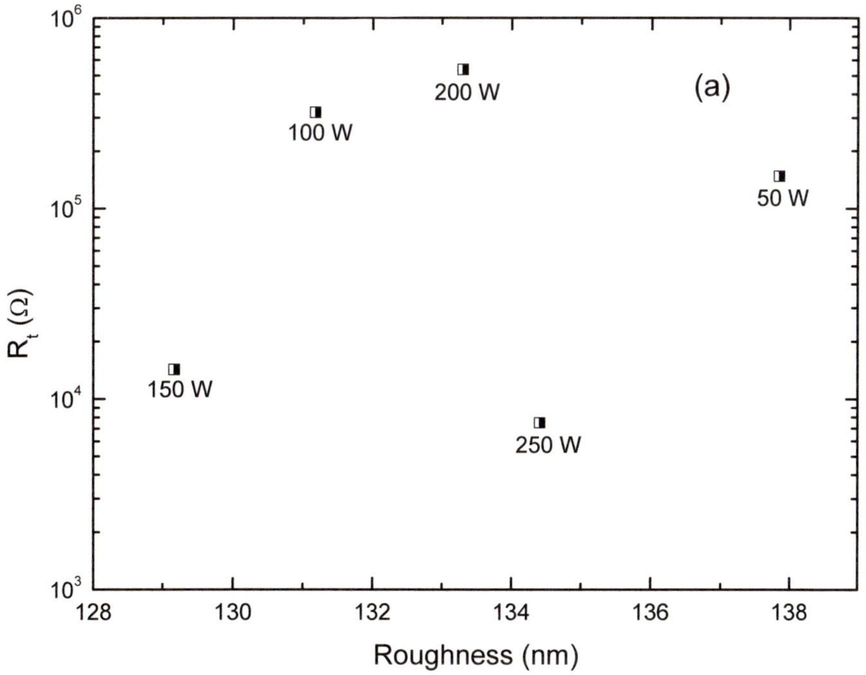

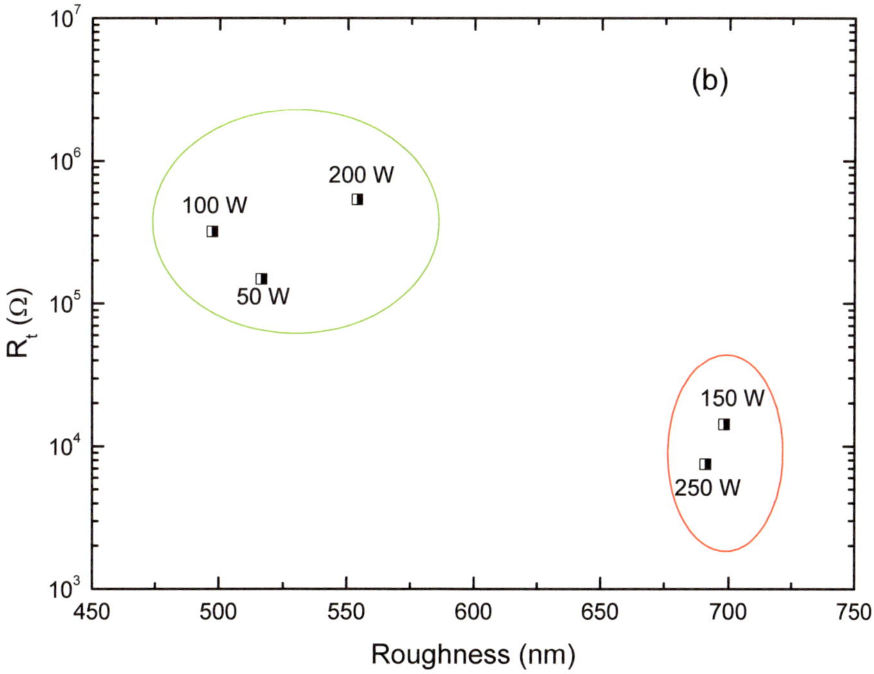

Fig. 19. Total resistance at t = 180 minutes as a function of roughness determined from AFM images (a) and from perfilometry profiles (b).

4. Conclusions

Molecular structure, chemical composition, morphology and topography of plasma deposited HMDSO films were strongly dependent on the plasma excitation power. The films are water repellent and present good adhesion to different substrates. Physical stability was a problem just for the layer prepared with the highest power in which thickness exceeded 10 μm, but the reduction in the deposition time would easily solve this drawback. All the other samples showed integrity even after aging in atmosphere conditions. The different reactivities of the hybrid organic-inorganic structure towards fluorine suggest it is a potential material for designing microelectromechanical devices. Moreover, films are transparent to visible light and highly resistant to oxygen attack, convenient properties as one considers the protection of a series of outdoor and space devices. In saline solutions, the most aggressive medium for steels, film deposition was observed to increase the corrosion resistance of the samples by more than 47.000 times. Differently of the results found in a previous work, the protection lasted until the end of the test, revealing an intact surface after the corrosion experiment. The optimum plasma treatment condition was considered to be the conducted with 200 W of power, since it provided the highest corrosion resistance for longer times. No intermediary oxide layer was grown in the metal surface and then all the improvements are ascribed to the HMDSO derived film. This opens the possibility of further improving the corrosion resistance of the carbon steel by creation of a dense oxide layer onto the metallic substrate prior to film deposition. New studies will be developed to evaluate this possibility. It should be finally mentioned that the films produced here are not exclusive to carbon steel protection since they adhere well to different substrates.

5. Acknowledgment

The authors would like to thank Brazilian agencies FAPESP and CNPq for financial support.

6. References

Biederman, H. e Osada, Y. 1992. *Plasma Polymerization Processes.* New York : Elsevier, 1992.

Bruce, R.L.; Lin, T.; Phaneuf, R.J.; Oehrleinb, G.S.; Bell, W.; Long, B.; Willson, C.G. 2010. Molecular structure effects on dry etching behavior of Si-containing resists in oxygen plasma. *J. Vac. Sci. Technol. B.* Jul 2010, Vol. 28, 4, pp. 751-757.

Chen, Q.; Zhang, D.; Tan, Z.; Wang, Z.; Liu, L.; Lu, J.Q. 2011. Thick benzocyclobutene etching using high density SF6/O2 plasmas. *J. Vac. Sci. Technol. B.* Jan 2011, Vol. 29, 1, pp. 011019-1 - 011019-6.

Choudhury, A.J.; Barve, S.A.; Chutia, J.; Pal, A.R.; Kishore, R.; Jagannath; Pande, M.; Patil, D.S. 2011. RF-PACVD of water repellent and protective HMDSO coatings on bell metal surfaces: Correlation between discharge parameters and film properties. *Appl. Surf. Sci.* Aug 2011, Vol. 257, pp. 8469-8477.

Choudhury, A.J.; Chutia, J.; Kakati, H.; Barve, S.A.; Pal, A.R.; Sarma, N.S.; Chowdhury, D.; Patil, D.S. Studies of radiofrequency plasma deposition of hexamethyldisiloxane films and their thermal stability and corrosion resistance behavior. *Vacuum.* Jun 2010, Vol. 84, 11.

Fracassi, F.; d'Agostino, R.; Palumbo, F.; Angelini, E.; Grassini, S.; Rosalbino, F. 2003. Application of plasma deposited organosilicon thin films for the corrosion protection of metals. *Surf. Coat. Technol.* Sep 2003, Vols. 174-175, pp. 107-111.

Gengenbach, T.R. e Griesser, H.J. 1999. Polymer. *Post-deposition ageing reactions differ markedly between plasma polymers deposited from siloxane and silazane monomers.* Aug 1999, Vol. 40, 18, pp. 5079-5094.

Guruvenket, S.; Azzi, M.; Li, D.; Szpunar, J.A.; Martinu, L.; Klemberg-Sapieha, J.E. 2010. Structural, mechanical, tribological, and corrosion properties of a-SiC:H coatings prepared by PECVD. *Surf. Coat. Technol.* Aug 2010, Vol. 204, 21-22, pp. 3358-3365.

Huang, Y.; Tian, X.; Lv, S.; Yang, S.; Fu, R.K.Y.; Chu, P.K.; Leng, J.; Li, Y. 2011. An undercutting model of atomic oxygen for multilayer silica/alumina films fabricated by plasma immersion implantation and deposition on polyimide. *Appl. Surf. Sci.* Aug 2011, Vol. 57, 21, pp. 9158-9163.

Koch, G.H.; Brongers, M.P.H.; Thompson, N.G.; Virmani, Y. P.; Payer, J.H. 2002. *Corrosion costs and preventive strategies in the United States.* Nace International. 2002. available at http://events.nace.org/publicaffairs/images_cocorr/ccsupp.pdf.

Lanford, W.A.; Rand, M.J. 1978. The hydrogen content of plasma-deposited silicon nitride. *J. App. Phys.* Apr 1978, Vol. 49, 4, pp. 2473-2477.

Mansfeld, F. 1981. Recording and analysis of AC impedance data for corrosion studies: I. Background and methods of analysis. *Corrosion.* May 1981, Vol. 37, 5, pp. 301-307.

Morosoff, N. 1990. *Plasma Deposition, Treatment and Etching of Polymers.* [ed.] R. d'Agostino. New York : Academic Press, 1990.

Muntasser; Z.M.; Al-Darbi, M.M.; Islami, M.R., 2002. *Corrosion.* 2002.

Quéré, D. 2005. Non-sticking drops. *Rep. Prog. Phys.* 68, Nov 2005, pp. 2495-2532.

Radhakrishnan, S.; Sonawane, N.; Siju, C.R. 2009. Epoxy powder coatings containing polyaniline for enhanced corrosion protection. *Progress in Organic Coatings.* Mar 2009, Vol. 64, 4, pp. 383-386.

Rao, A.P.; Rao, A.V. 2010. Modifying the surface energy and hydrophobicity of the low-density silica aerogels through the use of combinations of surface-modification agents. *Journal of Materials Science.* Jan 2010, Vol. 41, 1, pp. 51-63.

Ricci, M.; Dorier, J.; Hollestein, C.; Fayet, P. 2011. Influence of Argon and Nitrogen Admixture in HMDSO/O2 Plasmas onto Powder Formation. *Plasma Process. Polym.* Feb 2011, Vol. 8, 2, pp. 108-117.

Scheinmann, F. 1970. *An introduction to spectroscopic methods for the identification of organic compounds.* Oxford : Pergamon Press, 1970. Vol. 1.

Shin, A.S.; Shon, M.Y. 2010. Effects of coating thickness and surface treatment on the corrosion protection of diglycidyl ether bisphenol-A based epoxy coated carbon steel. *Journal of Industrial and Engineering Chemistry.* 16, Nov 2010, Vol. 6, pp. 884-890.

Stillahn, J.M., zhang, J., Fisherb, E.R. 2011. Surface interactions of SO2 and passivation chemistry during etching of Si and SiO2 in SF6/O2 plasmas. *J. Vac. Sci. Technol. B.* Jan 2011, Vol. 29, 1, pp. 011014-1 - 011014-10.

Tian, W.-C., Weigold, J.W., Pang, S.W. 2000. Comparison of Cl2 and F-based dry etching for high aspect ratio Si microstructures etched with an inductively coupled plasma source. *J. Vac. Sci. Technol. B.* Jul 2000, Vol. 18, 4, pp. 1890-1896.

Ul, C.V.; Roux, F.; Laporte, C.B.; Pastol, J.L.; Chausse, A. 2002. Hexamethyldisiloxane (HMDSO)-plasma-polymerised coatings as primer for iron corrosion protection: influence of RF bias. *J. Mater. Chem.* Jun 2002, Vol. 12, 8, pp. 2318-2324.

Ul, C.V.; Laporte, C.B.; Benissad, N.; Chausse, A.; Leprince, P.; Messina, R. 2000. Plasma-polymerized coatings using HMDSO precursor for iron protection. *Progress in Organic Coatings.* Feb 2000, Vol. 38, 1, pp. 9-15.

Yasuda, H. 1985. *Plasma Polymerization.* New York : Academic Press, 1985.

6

Corrosion Resistant Coatings Based on Organic-Inorganic Hybrids Reinforced by Carbon Nanotubes

Peter Hammer, Fábio C. dos Santos,
Bianca M. Cerrutti, Sandra H. Pulcinelli and Celso V. Santilli
*Instituto de Química, UNESP-Univ Estadual Paulista,
Brazil*

1. Introduction

Composite materials can be prepared by the combination of at least two different phases, a continuous and a dispersed one. If one of these phases is on the nanometer scale, than the material is called a nanocomposite. In this context organic-inorganic hybrids stand for a new class of materials formed by two distinct compounds, the polymeric and ceramic, resulting in a crosslinked amorphous nanocomposit exhibiting different properties than the initial phases. The appropriate choice of the type and proportion of ceramic and polymeric precursor results in unique properties of the hybrid material combining processability and flexibility of polymer compounds with thermal, chemical and mechanical stability of ceramic materials. (Zheludkevich et al., 2005; Wang & Bierwagen, 2009). This allows to design specific characteristics of the material for a wide range of applications. Different forms of organic-inorganic hybrids have been intensively studied due to their interesting mechanical, optical and thermal properties (Zheludkevich et al., 2005), resulting in a number of applications such as drug delivery systems, sensors, electrical and optical devices, catalysts and protective coatings (Landry et al., 1992; Sarmento et al., 2010). Due to their inert and nontoxic nature, hybrid coatings are considered among the existing corrosion inhibitors as the most promising candidates for environmentally compliant surface protection.

The preparation of organic-inorganic hybrid nanocomposits can be performed by different chemical and physical routes, resulting in a material in which the organic and inorganic phases interact by weak intermolecular forces (class I hybrid) or by establishment of covalent bonds (class II hybrid), forming in the latter case a dense cross-linked network (Zheludkewich et al., 2005). The most widely used method for the synthesis of hybrid materials is the sol-gel process, which does not require high temperatures or other extreme conditions, and provides homogenous, transparent and high purity material at low costs. The chemistry of the sol-gel process is based on the hydrolysis and polycondensation reactions of metal and semimetal alkoxides combined with the radical polymerization of the monomer of the organic precursor (Binkler & Scherer, 1990). To promote covalent bonds between the inorganic network and the organic compound often a coupling agent is used in the form of alkoxide groups modified by a vinyl binder. Inorganic materials prepared by the sol-gel route such as SiO_2, Al_2O_3, ZrO_2, etc., have relatively high hardness and thermal stability, but they posses also a high porosity (Vasconcelos at al., 2000; Nazeri at al. 1997; Maslski at al., 1999). Thus the inclusion of an organic component in the pores of the inorganic phase produces an ultra-light, dense, flexible and resistant material with low internal stress, suitable for a number of applications (Sanches & Lebeau, 2004). Among the various available organic substituted trialkoxysilanes, (3-methacryloxy propyltrimethoxysilane (MPTS) has been successfully used as coupling agent between silica and poly(methyl methacrylate) (PMMA) phases prepared from the co-reaction of methyl methacrylate (MMA), tetra-alkoxysilanes such as tetraethyl orthosilicate (TEOS), and tetramethyl orthosilicate (TMOS) (Innocenzi at al., 2003; Herreld at al., 2003; Sarmento et al., 2006; Han et al., 2007). It is evident that the properties of hybrid materials are not just the sum of individual contributions of its constituents, there is a synergism that depends also on the chemical nature of organic-inorganic interface and the size and morphology of the involved phases (José & Prado, 2005).

Metallic corrosion is induced by the chemical reactions between the metal surface and the environment, leading to an irreversible disintegration of the material under formation of oxides, hydroxides and salts. In the case of steel and aluminum, Cl^-, O_2, and H_2O species, in addition to electron transport, play important roles in the corrosion process (Ryan at al., 2002; Betova et al., 2002). The corrosion resistance is strongly weakened when metallic materials are subjected to a medium containing chloride or moisture environment, with a tendency to suffer localized corrosion. This may result result in the loss of aesthetic appearance and structural integrity, and can be accompanied by the release of potentially toxic ions, which can be hindered by protective coating (Tsutsumi at al., 2007; Bhattacharyya et al., 2008; Lopez at al., 2008; De Graeve at al., 2007). The most common corrosion prevention method is the hexavalent chromium-conversion, which forms insoluble trivalent chromium products by metal dissolution followed by precipitation of a passive layer of corrosion product. This passive layer on aluminum and steel surfaces should be capable of resisting the chemical attack, thus preventing further metal oxidation. However due to the carcinogenic effects of hexavalent chromium-containing species, environmental regulations have mandated the near term removal of Cr(VI)-containing compounds from corrosion inhibiting packages.

In this context, the design of new materials that act as a barrier against the diffusion of aggressive species have been widely investigated, primarily driven by the need to replace the corrosion inhibitors based on chromates. For this purpose, different types of coatings have been developed, both organic (paints), and inorganic (ceramic or conversion such as anodization) as well as a combination of organic and inorganic compounds. However, films based on polymeric materials have low thermal stability and poor adhesion to metal surfaces, while inorganic coatings suffer limitations due to micro cracks, porosity and high internal stress leading to adhesion problems and thickness limitations. Alternatively, the research on organic-inorganic hybrid materials deposited by dip or spin-coating on various substrates, mostly metals, has intensified in recent years due to their excellent anti-corrosion performance (Sarmento et al., 2010; Kim et al., 2009). A variety of different compositions including siloxane-methacrylate and ZrO_2-PMMA hybrid systems have been studied showing pronounced improvement of corrosion resistance of stainless steel alloys against corrosion under acidic conditions and in the presence of high concentrations of Cl^- ions (Messaddeq et al., 1999; Meteroke et al., 2002; Zandri-Zand et al. 2005; Zheng & Li, 2010). This is a consequence of their highly interconnected structure of ramified siloxane cross-link nodes interconnected by short polymeric chains forming a chemically inert barrier, which prevents the penetration of species that initiate corrosive processes (Sarmento et al., 2010). Detailed analysis of the system have shown that inorganic phase has an important role in promoting the adhesion between the film and metal substrate, while the organic phase hermetically seals the film structure (Sarmento at al., 2010). In addition, hybrid films exhibit hydrophobic character and as a consequence of their low intrinsic stress they can be prepared crack-free even with a thickness of several micrometers (Messaddeq et al., 1999; Sarmento at al., 2010). Several other aspects favor the use of these materials, among which can be highlighted the mild conditions of the sol-gel synthesis, the low costs of preparation, environmental compatibility and also the low temperatures needed to cure the coatings. Consequently, organic-inorganic hybrid films prepared via sol-gel process posses a great potential for large-scale application as protective coatings of metallic surfaces.

Recently, several studies reported on the improvement of the passivation character of hybrid coatings by the addition of species that act as corrosion inhibitors. In particular cerium III and IV ions has been shown to satisfy several basic requirements expected from a corrosion inhibitor, for example, to increase the degree of polycondensation of the amorphous network, to form insoluble nontoxic oxides and hydroxides, besides to be relatively cheap and easy to handle (Pepe et al., 2004). The effects of cerium ions in the structure of the hybrids deposited by dip-coating has been investigated by several authors (Pepe et al., 2004; Suegama et al., 2009), including our group (Hammer at al., 2010), showing from the analysis of nuclear magnetic resonance and X-ray photoelectron spectroscopy data the active role of Ce(IV) on the densification of the polysiloxane films, i.e., the finding that free radical reactions between neighboring silanol and ethanol groups, induced by the reduction of Ce(IV), increases the connectivity of the hybrid network, resulting in elevated corrosion protection efficiency of the coatings.

Another important issue is the thermal and mechanical stability of hybrid films. Composite materials like polymer and hybrid systems containing carbon nanotubes (CNTs) are gaining increasing space as a promising class of advanced materials (Kim at al., 2009). The high thermal and mechanical stability and chemical inertness of carbon nanotubes (Khare at al., 2005) makes them a very interesting candidate for the improvement the properties of horganic-inorganic hybrid materials. Recently, the incorporation of carbon nanotubes in polymers and hybrids has shown excellent results in terms of improvement of their mechanical stability (Eder, 2010; Kim at al., 2009). The inclusion of only 0.1 to 1.0 at.% of functionalized (Voiry at al., 2011) or doped (Droppa Jr., 2001) carbon nanotubes is sufficient to improve considerably the mechanical properties and to increases the electrical and thermal conductivity of nanocomposites (Eder, 2010). Applications of these highly resistant and lightweight materials are in fields of electronics, energy, chemical industry and construction including, among others, flexible polymers, epoxy resins, refractory materials and nanostructured concrete. Despite the increasing interest to reinforce different classes of materials by carbon nanotubes, the area of hybrid materials is still not much explored. To improve the mechanical properties of organic-inorganic, carbon nanotubes functionalized by hydroxyl, amide or sulphate groups are of particular interest since they are able to promote covalent bonds with siloxane groups of the network. Thus from the conjugation of CNTs with organic-inorganic hybrid it is expected to obtain nanocomposite coatings that combine high anti-corrosion efficiency with elevated mechanical resistance. In addition, the metallic conduction of multi-wall nanotubes (MWCNTs) allows an increase in conductivity of the hybrids by several orders of magnitude. This effect in combination with an organic phase based on conjugated polymers (Sanches & Lebeau, 2004) can be exploited for the development of transparent conductive hybrid coatings with optimized passivating character, a material with the potential for numerous applications.

Considering the great potential and versatility of this new class of composite materials this work focuses on the investigation of sol-gel siloxane-polymethyl methacrylate hybrids conjugated with functionalized carbon nanotubes used as corrosion resistant coatings on carbon steel. Based on the correlation of structural data, obtained for hybrids containing different amounts of the organic phase, with their corrosion protection efficiency, the aim of this work was to identify the most suitable matrix for the incorporation of functionalized carbon nanotubes to reinforce the hybrid structure without degrading the corrosion protection character of the coatings. The structural features of the siloxane-PMMA hybrids were studied as a function of the PMMA fraction and carbon nanotubes concentration and ^{29}Si nuclear magnetic resonance (NMR), X-ray photoelectron spectroscopy (XPS) and thermogravimetric analysis (TGA) while the corrosion protection efficiency was investigated by electrochemical impedance spectroscopy (EIS) and potentiodynamic polarization curves, after immersion of the samples in a acidic and neutral saline solution.

The low carbon alloy steel was chosen as substrate since it combines properties such as high hardness, high toughness and high flexibility, besides being relatively inexpensive, easy to manufacture and having many functional applications such as automotive sheet metal, structural shapes, plates for production pipes, construction material, tin cans etc. However, corrosion is still a great obstacle when it comes to its durability, suffering severe corrosion when in contact with even low amounts of chloride ions and acid solutions.

2. Experimental procedure

2.1 Materials

The following reagents were used as received and are available commercially: tetraethylorthosilicate (TEOS, Aldrich), 3-methacryloxy propyltrimethoxy-silane (MPTS, Fluka), methyl methacrylate (MMA, Fluka), ethanol (Mallinckrodt) and tetrahydrofuran (THF, Mallinckrodt). The MMA was pre-distilled to remove the polymerization inhibitor (hydroquinone) and possible impurities. The benzoyl peroxide (BPO, Reagan) was recrystallized in ethanol solution. For doping experiments multi wall carbon nanotubes (Nanocyl) were used.

2.2 Preparation of the hybrids

The siloxane-polymethyl methacrylate hybrids were prepared using the sol-gel route in three stages: In the first step the monomer (MMA) and the alkoxide modified with a methacrylate group (MPTS) were polymerized in THF using the thermal initiator (BPO). The reaction was carried out during 2 hours at temperature of 70 °C under constant stirring in a reflux flask. In the second step, the inorganic component of the hybrid was prepared by hydrolysis and condensation of silicon alkoxide (TEOS). The hydrolysis of TEOS was carried in an ethanol solution by addition of acidified water (pH 1) using nitric acid. After 1 h of reaction at 25 °C under constant stirring in a closed flask, in a final step, the inorganic component was added into the reflux flask containing the organic solution. At this stage the modified MPTS is partially hydrolyzed and condensed with the TEOS, forming a homogeneous and transparent sol used for the deposition of the films by dip-coating. The following thermal cure 150 °C ensured a high degree of polycondensation of the hybrid structure. To study the influence of the proportion of the organic component on the film structure and electrochemical properties, the hybrids were prepared with the following MMA to MPTS molar ratios: 2, 4, 6, 8 and 10, referred to as M2, M4, M6, M8 and M10, respectively. The other optimized molar ratios were kept constant: $H_2O/Si = 3.5$, etanol/H_2O = 0.5, BPO/MMA = 0.01 and TEOS/MPTS = 2.0 (Sarmento et al., 2010).

2.3 Preparation of hybrids modified by carbon nanotubes

One of the hybrid samples with a higher fraction of silica and good anti-corrosion performance (M4) was chosen for the incorporation of functionalized carbon nanotubes (CNTs). The CNTs were functionalized by carboxyl groups in a standard oxidation procedure using a mixture of sulfuric and nitric acid (3:1) under reflux during 4 h at 70 °C. After dispersion in 0,5 L of deionized water and a short ultrasonic treatment the CNTs were filtered and annealed during 12 h at 200 °C. Finally, the functionalized carbon nanotubes were dispersed in a solution of ethanol containing Nafion® 5% and then added to the inorganic component of the hybrid. To study the influence of the CNTs content on film structure and electrochemical properties of the M4 sample, the following $[C_{CNT}]/[C_{CNT}+ Si]$ atomic concentrations were used: 0.1%, 1.0% and 5.0%, referred to as M4_01, M4_1 and M4_5, respectively. The aspect of the samples M4 and M4_5 obtained after 5 days of gelatinization are shown in Figure 1. Although the sample containing CNTs showed a dark coloration, the films deposited on steel hat the same colorless and transparent aspect as the undoped coatings.

Fig. 1. Hybrid gel samples prepared at MMA/MPTS ratio of 4, M4 without CNTs (transparent) and M4_5, containing CNTs (dark), obtained after 5 days of drying at 10 °C (see text for details).

2.4 Preparation of the substrates and film deposition

For electrochemical corrosion tests and XPS measurements, coated A1010 carbon steel substrates (25 mm x 20 mm x 5 mm) were used, having a nominal composition (wt%) of C = 0.08 to 0.13%, Mn = 0.3 to 0.6%, P = 0.04% max and S max = 0.05%, with the balance consisting of Fe. The deposition of hybrid films was performed by dip-coating process using a dip-coater (Microchemistry - MQCTL2000MP), optimized at a rate of 14 cm min^{-1}, with 1 min of immersion and air-drying during 10 min at room temperature. This procedure was performed three times for each sample. Then the coated substrates were heated at 60 °C for 24 h. Finally, the samples were subjected to further heat treatment at 160 °C during 3 h. The thermal treatment at 60 °C favors the radical polymerization (initiated by BPO) of methacrylate groups of MPTS and of MMA monomers in the film and the cure at 160 °C was carried out to ensure the entire removal of all solvents, to complete the polymerization process of organic compounds and to densify the film. Homogeneous, transparent, crack-free coatings were produced. The coating thickness was determined by profilometry (Talystep, Taylor & Hobson). The adhesion of the films to the steel surface was carried out using a PosiTest Pull-Off Adhesion Tester (De Felsko), testing the coating´s strength of adhesion to the substrate by determining the tensile pull-off force of detaching.

2.5 Characterization methods

Unsupported hybrid films for NMR and TGA analysis were obtained by drying the sols in a Petri dish at 60 °C for 24 h. After the curing process, at 160 °C for 3 h, the films were detached from the dish and milled to obtain a powder. To obtain structural information solid-state ^{29}Si magic-angle spinning nuclear magnetic resonance (MAS-NMR) spectra were recorded for powder samples using a VARIAN spectrometer operating at 300 MHz and 7.05 T. The Larmor frequency for ^{29}Si was 59.59 Hz. The spectra were obtained from the Fourier transformation of the free induction decays (FID), following a single $\pi/2$ excitation pulse and a dead time of 2 s. Chemical shifts were referenced to tetramethylsilane (TMS), used as external standard. Proton decoupling was always used during spectra acquisition. Because of the high sensitivity of the ^{29}Si NMR measurements, the uncertainty in the chemical shift values was less than 0.2 ppm.

The TGA curves of powder samples were recorded using a SDT Q600 (TA Instruments) thermal analysis system, with nitrogen as purge gas at a flow rate of 70 ml min^{-1}. About 7 mg samples of hybrids were placed into platinum crucibles, and heated at a rate of 10 °C min^{-1} up to 600 °C.

The XPS analysis was carried out at a pressure of less than 10^{-6} Pa using a commercial spectrometer (UNI-SPECS UHV) to verify the changes in of the local bonding structure of samples containing different amounts of polymeric phase and carbon nanotubes. The Mg Kα line was used (hν= 1253.6 eV) and the analyzer pass energy for the high-resolution spectra was set to 10 eV. The inelastic background of the C 1s, O 1s and Si 2p electron core-level spectra was subtracted using Shirley's method. Due to charging of the samples, the binding energies of the spectra were corrected using the hydrocarbon component of adventitious carbon fixed at 285.0 eV. The composition of the near surface region was determined with an accuracy of ±10% from the ratio of the relative peak areas corrected by Scofield's sensitivity factors of the corresponding elements. The spectra were fitted without placing constraints using multiple Voigt profiles. The width at half maximum (FWHM) varied between 1.0 and 2.0 eV and the accuracy of the peak positions was ±0.1 eV.

The polarization resistance of coated and uncoated steel samples was evaluated by means of electrochemical measurements carried out at 25 °C in 400 mL of naturally aerated and unstirred 0.05 mol L^{-1} H_2SO_4 + 0.05 mol L^{-1} NaCl solution and in neutral 3.5% NaCl solution. An Ag/AgCl/KCl_{sat} electrode, connected to the working solution through a Luggin capillary, was used as reference, and a Pt grid as the auxiliary electrode. The working electrode was mounted in an EG&G electrochemical flat cell, exposing a geometric area of 1 cm^2 to the solution. This area is generally different of the actual area exposed to the solution, since it depends on the electrolyte penetration, surface roughness and on the many defects present in the layer (Suegama et al., 2006).

Polarization curves were recorded for all samples using a potentiostat/galvanostat (EG&G Parc-273), over the potential range -150 mV to +1000 mV versus the open circuit potential, E_{OC}, referred to the Ag/AgCl/KCl_{sat} electrode, at a scan rate of 0.167 mV s^{-1}. The polarization curves were recorded after 4 hours immersion in the electrolyte. The corrosion potentials and apparent corrosion current density values were directly estimated from classical *E* (mV) *vs. log i* (A cm^{-2}) curve. The Stern-Geary equation (Stern & Gaery, 2006) was not utilized, since not all conditions required for its use were fulfilled in the system studied. The Stern-Geary relationship can only be applied for general corroded surfaces when the conditions necessary to deduce the Buthler-Volmer equation are valid. This means that the polarization resistance should be equal to the charge transfer resistance of the corrosion reaction (Feliu et al., 1998). However, the polarization resistance can be comprised of diffusion resistance, adsorption resistance, ohmic resistance and other resistances, as a result of which the conditions needed for using the Stern-Geary equation are often not fulfilled. Therefore, discussion of the corrosion mechanism based on polarization curves was ruled out. For that reason, the polarization curves were qualitatively analyzed and only used to compare the coated materials with different formulation. To guarantee the reproducibility of the results all electrochemical measurements were performed in duplicate.

In another series of experiments, open circuit potential (EOC) and electrochemical impedance spectroscopy (EIS) measurements were performed for M2, M4, M8 and M10 samples after 1 day of immersion in acidic NaCl solutions. For the M8 sample, considered to be the hybrid coating with the best performance, the EIS studies were conducted for up to 18 days of immersion in saline solution. The EIS measurements were performed using a Potentiostat/Galvanostat EG&G Parc-273 and a Frequency Response Analyzer Solartron-SI1255 coupled to a computer. The EIS tests were performed applying 10 mV (rms) to the E_{OC} value, starting from 10^5 to $5x10^{-3}$ Hz with 10 points/decade. A 0.1 μF capacitor was connected to a platinum wire and a reference electrode to avoid phase shifting at high frequencies and noise at low frequencies. For all samples, E_{OC} was measured for 2 h, time enough to stabilize the potential value. Afterwards, the E_{OC} value was applied for 15 min with simultaneous measuring of the current, which stabilized during this time, and the impedance spectra were immediately recorded in 3.5% NaCl and in acidic chloride-containing solution. The experiments were performed in duplicate.

Finally, atomic force microscopy (AFM) measurements were carried out using the standard tapping mode of the Aglient 5500 instrument, the morphology of the surface as investigated by field emission guns scanning electron microscopy (FEG-SEM) in the secondary electrons mode at 3 kV (JEOL 71500F) and the contact angle measurements were carried out by the sessile drop method using an OCA-20 Contact Angle System (DataPhysics Instruments).

3. Results and discussion

3.1 Pure hybrids

3.1.1 Structural analysis

AFM images of the hybrid coating showed a very smooth surface with an RMS roughness of less than 0.2 nm (M2). Inspections by optical microscopy and scanning electron microscopy (Fig. 2) confirmed that the transparent films are homogeneous and defect-free. The film thickness, determined by profilometry, was about 1.5 μm for the M2 coating. The tensile pull-off force of detaching was for all coatings higher than the 8 MPa limit of the PosiTest Pull-Off Adhesion Tester.

Figure 3 shows ^{29}Si NMR spectra for unsupported siloxane-polymethyl methacrylate hybrids prepared with MMA/MPTS molar ratios of 2, 4, 8 and 10 (M2, M4, M8 and M10). The main spectral feature of the samples is the broad resonance with peaks at -59 and -65 ppm, assigned to T^2 ($-CH_2Si(OSi)_2(OR)$, R = H or CH_3) and T^3 ($-CH_2Si(OSi)_3$) units, which are related to the polycondensation product of MPTS (Saravanamuttu et al., 1998; Tadanaga et al., 2000). Resonances related to TEOS polycondensation products are observed, at approximately -92, -102 and -109 ppm, which correspond to Q^2 (Si(OSi)2(OR)2), Q^3 ($Si(OSi)_3(OR)$) and Q^4 ($Si(OSi)_4$) species, respectively (Fedrizzi at al., 2001; Saravanamuttu et al., 1998; Tadanaga et al., 2000). The absence of monomer species T^0 (-CH2Si(OR)3) and Q^0 ($Si(OR)_4$) is in agreement with the cluster-cluster condensation mechanism, which is expected under strongly acid conditions (Han at al., 2000). The proportions of T^J and Q^J species present in the hybrid nanocomposit were extracted from the spectra by a peak fitting procedure used to determine the relative peak area of each species. As expected, the intensity ratio of Q^J/T^J species was found to be approximately 2, in agreement with the proportions of the TEOS and MPTS precursors used in the sample synthesis, which differed only in the fraction of the organic phase. The degree of condensation ($\%C_d$) of the inorganic phase in the unsupported polysiloxane hybrid was calculated from the proportions of each T^J and Q^J species according to the following equation (Saravanamuttu et al. 1998):

$$\%C_d = [(T^1 + 2T^2 + 3T^3)/3 + (Q^1 + 2Q^2 + 3Q^3 + 4Q^4)/4] \times 100 \qquad (1)$$

The degree of condensation obtained for M2, M4, M8 and M10 hybrids were 80.9±0.5%, 79.7±0.5%, 83.9±0.5% and 75.8±0.5%, respectively. This result shows that with exception of M10 all samples have a high and almost identical degree of condensation and thus a similar bonding structure of the inorganic phase. This is not surprising since all hybrids were prepared at a fixed TEOS/MPTS ratio, restricting the variability of the silica network. The slightly higher degree of condensation of M8 is due to the increase of both the T^3/T^2 and Q^4/Q^3 the ratio, suggesting that heterocondensation occurs between both the MPTS and TEOS and the TEOS hydrolyzed species. The pure connectivity of M10 is evident (increase of T^1 and Q^2), and probably related to the excess of polymeric phase hindering the cross-linking of silica. Similar connectivity values as observed for M8 were found for highly corrosion resistant polysiloxane hybrids having equal TEOS/MPTS molar ratio, however without inclusion of the PMMA phase (Sarmento et al., 2010).

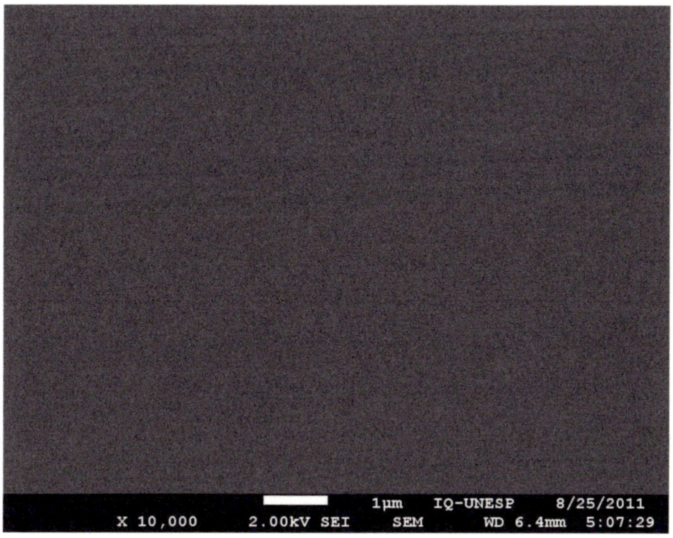

Fig. 2. Representative FEG-SEM image of the featureless surface of hybrid films (M4).

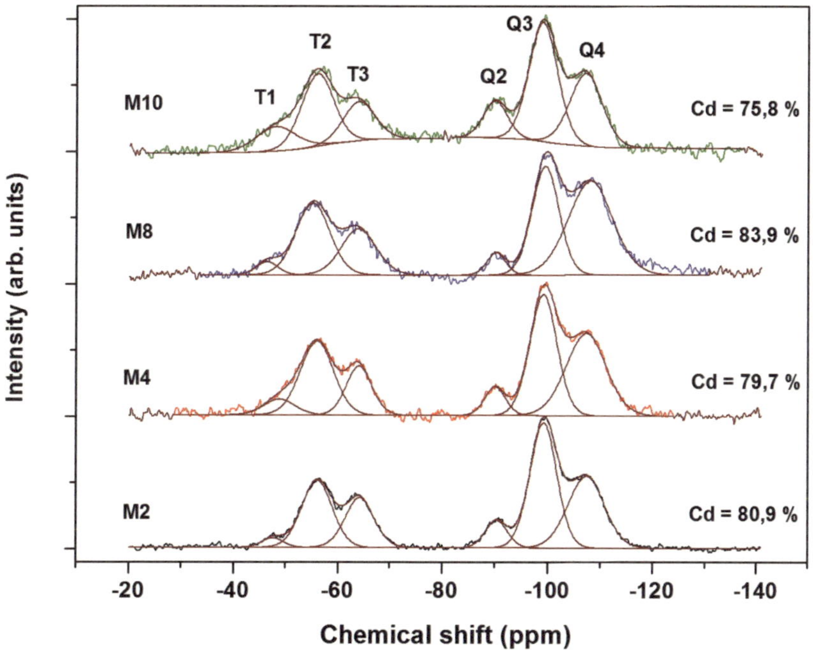

Fig. 3. ^{29}Si NMR spectra of hybrids prepared with MMA/MPTS ratios of 2 (M2), 4 (M4), 8 (M8) and 10 (M10).

The polymerization of the organic moieties was confirmed by the TGA curves and the differential weight loss (DTG) of unsupported siloxane-polymethyl methacrylate hybrids, shown in Figure 4. Since the sample were cured at 160° C the small weight loss observed below 110 °C is attributed to the water vapor adsorption. The DTG curves indicate the existence of three degradation stages. The first two events at about 240 °C and 390 °C correspond to the radical depolymerization of the organic polymer segments of the hybrid. The event observed at about 240 °C might be ascribed to scissions of the chain initiated from the vinylidene end, and the one in the range of 390 °C - 410 °C to random scission within the polymer chain (Han at al., 2000). The weak degradation stage above 450 °C, can be attributed to the dehydration of silanol groups corresponding to Q^2 and Q^3 species present in the SiO_2 network (Han at al., 2000). With exception of the M2 hybrid, showing the highest thermal stability of about 410 °C, the temperature corresponding to each degradation step remains essentially invariant, suggesting that all other hybrids have a similar silica backbone which is not essentially affected by the increasing presence of the polymerization phase. It is interesting to note that the less stable step, involving the scissions of head-to-head linkages in the PMMA homopolymer, is absent in the case of the unsupported hybrid films. This is another indication that the prepared hybrids possess a much higher thermal stability than the PMMA homopolymer.

It is important to note that structural characteristics of thin film produced by dip-coating and the thick bulk film formed in Petri dish, used for NMR and TGA analysis, were found be equal by XPS analysis, thus guarantying a consistent interpretation of structural and electrochemical data.

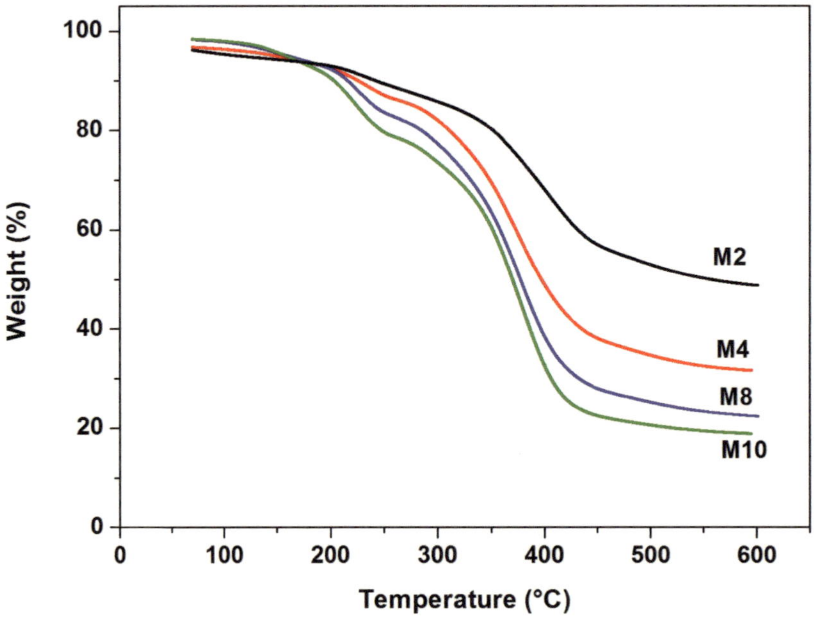

Fig. 4. TGA and DTG (insert) curves of hybrids prepared with MMA/MPTS ratios of 2 (M2), 4 (M4), 8 (M8) and 10 (M10).

X-ray photoelectron spectroscopy was used to obtain further information on the bonding structure and composition of the hybrid. The quantitative analysis showed that the detected elemental concentrations is in agreement with the nominal composition of the samples, showing an increase of the atomic percentage of carbon from 51.9 at.% (M4) to 60.7 at.% (M10) accompanied by a reduction of oxygen content from 37.3.% (M4) to 33.1 at.% (M10) and of silicon from 10.8.% (M4) to 6.2 at.% (M10).

The structural evolution of the samples, with increasing TEOS/MPTS ratio, was evaluated from the fitted high-resolution C 1s, O 1s and Si 2p core level spectra, displayed in Figure 5.

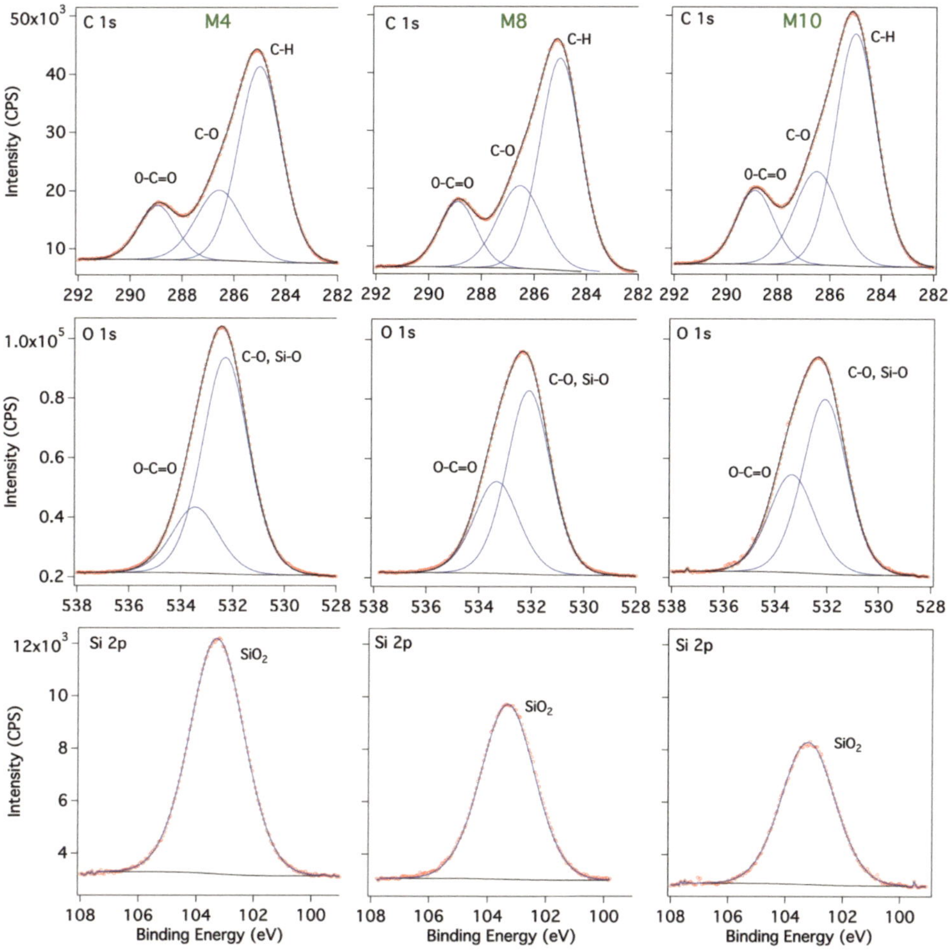

Fig. 5. Fitted XPS, C 1s, O 1s and Si 2p core level spectra of hybrids prepared with MMA/MPTS ratios of 4 (M4), 8 (M8) and 10 (M10).

The C 1s spectra show for all hybrids the presence of three different structural components of the organic phase, corresponding to hydrocarbon (C-H, at 285.0 eV), ether (C-O, at 286.6 eV) and ester (O-C=O, at 289.1 eV). It can be observed that with increasing fraction of the polymeric phase in the samples (M4, M8, M10) the intensity o the spectrum increases while intensity ratio of the sub-peaks I(C-H) : I(C-O) : I(O-C=O) remains constant, reflecting the preservation of the structural units of PMMA in the hybrid. The O 1s spectra were fitted with two components: the main peak located at 532.6 eV is related to C-O and Si-O bonds, while the small component at about 534.0 eV can be assigned to a ester groups (already observed in the C 1s spectrum). The decrease of the intensity of the O 1s spectra and the increase of the ester component evidence the decreasing proportion of silica in the samples. Similar evolution was observed for the Si 2p spectra, showing the continuous decrease of its intensity upon increasing fraction of the polymeric phase.

3.1.2 Corrosion analysis

To investigate the corrosion protection performance of hybrids containing different fractions of polymeric phase electrochemical impedance spectroscopy curves were recorded after immersion periods of up to 3 days in acidic solution and up to 18 days in saline environment. Figure 6 displays the complex plane impedance and the Bode plots (log $|Z|$ and θ vs. log f) for the coated samples (M2, M4, M8, M10) and bare steel after exposure of 1 day in neutral saline solution and after a period of 3 days in acidic NaCl solutions. Compared to bare steel, the coated samples show up to 5 orders of magnitude higher impedance with best performance observed for the M8 hybrid coating, reaching about 1 GΩ in both environments. This value is about two orders of magnitude higher than the highest impedance reported for polysiloxane hybrids prepared with equal TEOS/MPTS molar ratio of 2 but without inclusion of MMA (Sarmento et al., 2010). The phase angle dependence shows a broad band extending from 5 mHz to 100 kHz, with θ values close to 90°, apparently with only one time constant close to 10 kHz. Tests with different equivalent circuits models have shown, however, a poorly defined time constant in the low frequency region. No significant change of the curve characteristics was observed after 3 days of exposure in acidic and 18 days in saline solution (Fig. 7), confirming the excellent performance of this coating. A similar high stability in both environments was observed for the M2 coating, having, however, two orders of magnitude lower corrosion resistance and a narrower θ frequency band. In contrast, for M4 and M10 coatings two well defined time constants were observed after 3 days of immersion in acid environment, with the low frequency time constant appearing at the position of the time constant of bare steel.

The electrical equivalent circuit used for fitting impedance experimental data for hybrid films in both environments is displayed in Figure 8. According to the standard interpretation used for coated metals, the time constant in the high frequency region (HF) is attributed to the hybrid coating properties (coating capacitance (CPE_1) in parallel with its resistance (R_1)), while that in the low frequency region (LF) is related to the properties of the oxidized steel at the film/steel interface (oxide capacitance (CPE_2) in parallel with its resistance (R_2)). The formation of this passive film is a result of the reaction with the solution, which penetrates the coatings after extended immersion periods. The obtained model parameters are displayed in Table 1 and Table 2 for samples immersed in neutral saline solution and acidic environment, respectively. Compared with the M8 coating a strong decrease of resistances R_1 and R_2 was observed for M2 and M4 in saline medium, while the capacitances remain essentially unchanged (Table 1, Fig. 5a). The long-term exposure of M8 confirmed the superiority of this coating, showing only a small reduction of R_1 after 18 days in saline environment (Table 1, Fig. 6). After 3 days of exposure in acid environment M8 maintained its inert character while the M4 coating showed signals of beginning degradation (Table 2, Fig. 5b). The appearance of the LF time constant and the decrease of the capacitive character of the M4 and M10 coatings is indicative the beginning film failure after prolonged exposure in aggressive medium (Fig. 6c2). This is associated to electrolyte uptake, which increases the coating conductivity by the permeation of ionic species. For coatings M2 and M10 a corrosion process controlled by diffusion (Warburg element) cannot be excluded. Inspections by optical microscopy of these films revealed the formation of localized pits and the presence of iron species on the surface, detected by XPS.

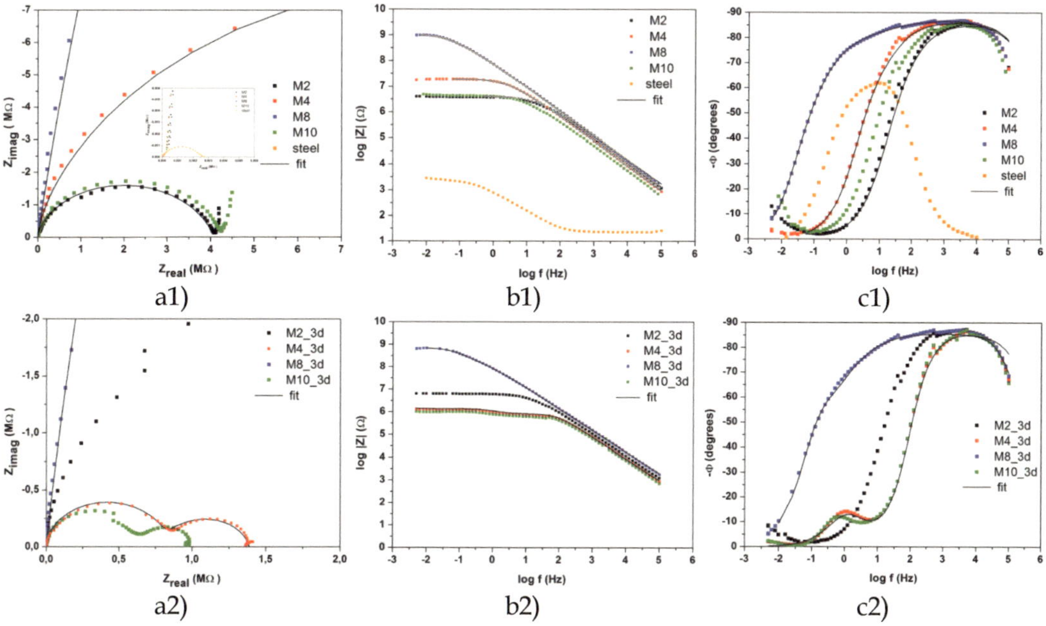

Fig. 6. Complex plane impedance (a), modulus impedance (b) and phase (c) plots of bare carbon steel and hybrids prepared with MMA/MPTS ratios of 2 (M2), 4 (M4), 8 (M8) and 10 (M10) after exposure of 1 day and 3 days in (1) unstirred and naturally aerated 3.5 % NaCl and (2) 0.05 mol L^{-1} NaCl + 0.05 mol L^{-1} H_2SO_4 solutions, respectively. (The small discontinuities of θ vs. log f mark the change of the measurement range of the instrument.)

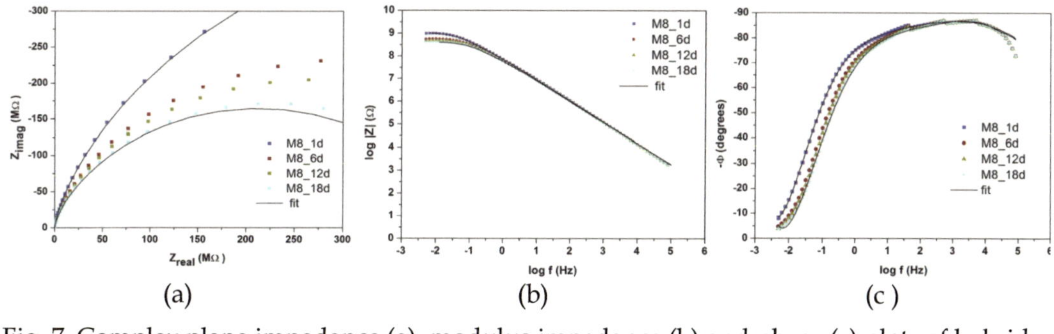

Fig. 7. Complex plane impedance (a), modulus impedance (b) and phase (c) plots of hybrids prepared with MMA/ MPTS ratios of 8 (M8) after exposure of up to 18 day in unstirred and naturally aerated 3.5% NaCl solution. (The small discontinuities of θ vs. log f mark the change of the measurement range of the instrument.)

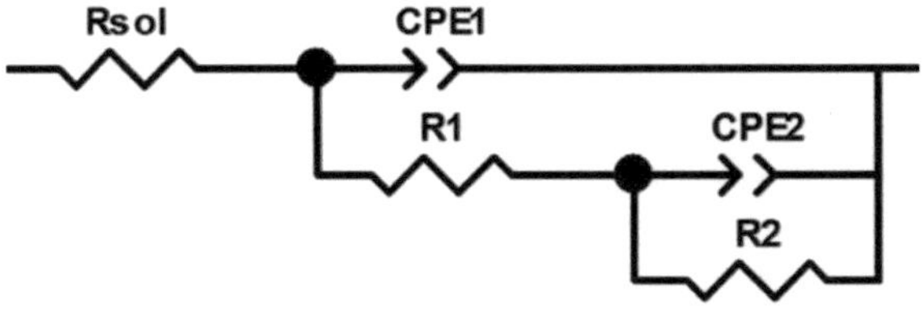

Fig. 8. Electrical equivalent circuit used for fitting impedance experimental data for hybrid films in 3.5 % NaCl and 0.05 mol L^{-1} H_2SO_4 + 0.05 mol L^{-1} NaCl solutions.

Sample	M8_1d	M8_18d	M4_1d	M2_1d
χ^2 (10^{-3})	9.8	6.4	9.7	9.5
R_{sol} (Ω cm^2)	22.7 (9.8)*	29.5 (4.4)	22.5 (7.8)	24.6 (7.8)
CPE_1 (10^{-3} μF cm^{-2} $s^{\alpha-1}$)	1.52 (28.9)	1.74 (4.2)	6.13 (16.3)	1.37 (16.9)
n_1	0.67 (7.6)	0.65 (4.5)	0.73 (8.1)	0.67 (7.4)
R_1 (10^6 Ω cm^2)	812 (12.5)	419 (2.9)	12.9 (11.6)	2.08 (16.5)
CPE_2 (10^{-3} μF cm^{-2} $s^{\alpha-1}$)	2.37 (3.4)	2.06 (2.4)	3.01 (3.1)	2.42 (4.9)
n_2	0.93 (6.4)	0.95 (0.25)	0.96(6.4)	0.95 (1.1)
R_2 (10^6 Ω cm^2)	305 (31.2)	26.9 (15.7)	6.72 (19.9)	2.01 (15.1)

*Error (%)

Table 1. Parameters of electrical equivalent circuit for sample M8, M4 and M2 after 1 day and 18 days (M8) immersion in neutral 3.5% NaCl aqueous solution.

Figure 9 displays the potentiodynamic polarization curves for uncoated, M2, M4, M8 and M10 coated carbon steel, recorded after 2 h of immersion in neutral NaCl solutions. As expected the polarization curve of the M8 coating, having the highest degree of condensation of the inorganic phase (highest degrees of condensation, see Fig. 3), exhibits the lowest current density of approximately 10^{-10} Acm^{-2}, a value about five orders of magnitude lower than that observed for bare steel. To our best knowledge such excellent anti-corrosion performance at similar experimental conditions was not yet reported in the literature for organic-inorganic hybrid films. The curve characteristics of M4 and M2 are quite similar, the latter showing slightly higher current densities. The negative open circuit potential of sample M10, found close to the value of bare steel, is most probably related to its less cross-linked network (see Fig. 3) suggesting a more open structure of this film. For all tested coatings no breakdown of the coating was observed in the positive potential branch even at overpotential of 1 V. The results obtained in acidic medium are very similar to those observed in the saline solution (not shown). According to the electrochemical results, the performance against corrosion can be established in the following order: M8 > M2 > M4 > M10 > steel. This result is in agreement with the sequence of increasing network connectivity values found by NMR.

Sample	M8_3d	M4_3d
χ^2 (10^{-3})	9.7	8.8
R_{sol} (Ω cm^2)	22.7 (9.8)*	29.4 (6.8)
CPE_1 (10^{-3} μF cm^{-2} s$^{\alpha-1}$)	1.56 (4.2)	0.04 (7.2)
n_1	0.65 (4.5)	0.86 (3.8)
R_1 (10^6 Ω cm^2)	754 (5.4)	0.56 (2.9)
CPE_2 (10^{-3} μF cm^{-2} s$^{\alpha-1}$)	1.52 (2.5)	2.60 (2.6)
n_2	0.95 (0.3)	0.97 (0.3)
R_2 (10^6 Ω cm^2)	49.1 (16.2)	0.83 (1.1)

*Error (%)

Table 2. Parameters of electrical equivalent circuit for sample M8 and M4 after 3 days immersion in 0.05 mol L^{-1} NaCl + 0.05 mol L^{-1} H_2SO_4 solution.

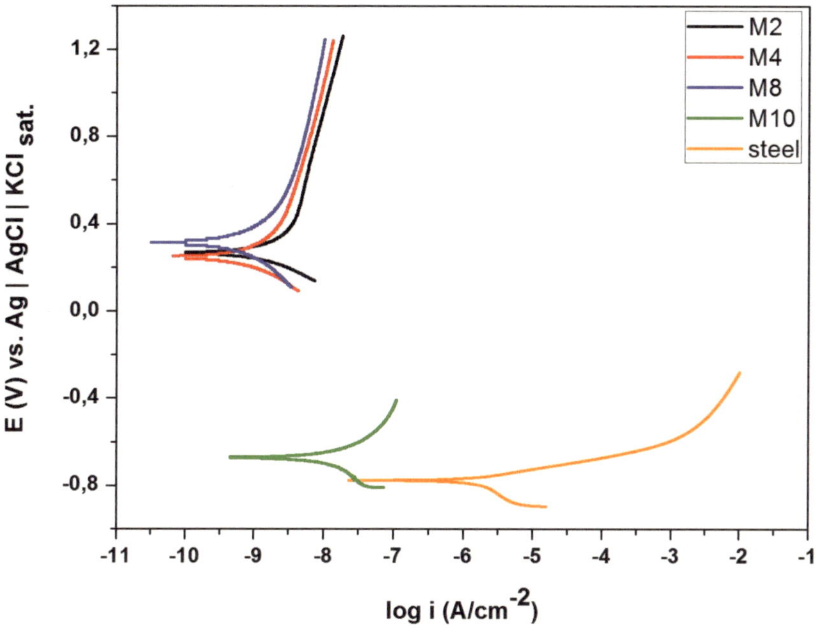

Fig. 9. Potentiodynamic polarization curves of bare carbon steel and hybrids prepared with MMA/MPTS ratios of 2 (M2), 4 (M4), 8 (M8) and 10 (M10) recorded after 2 h immersion in unstirred and naturally aerated 3.5 % NaCl solution.

3.2 Hybrids modified by carbon nanotubes

3.2.1 Structural analysis

For the experiments with samples containing functionalized carbon nanotubes the M4 hybrid was chosen due its higher fraction of silica and good anti-corrosion performance. Similar as in the case of pure hybrid films, AFM images of the M4_5 film, with a $[C_{CNTs}]/[C_{CNT}+ Si]$ molar concentration of 5%, have shown a very smooth surface (R_{RMS} =

0.16 nm) without features which could be related to the presence of carbon nanotubes on the film surface (Fig. 10a), expected for MWCNTs at a scale of 10 – 20 nm. A featureless surface on the nanometer scale was also confirmed by FEG-SEM, as can be observed in Figure 10b. TThis indicates a very good dispersion of functionalized carbon nanotubes in the hybrid matrix. The film thickness, determined by profilometry, incrased from 1.5 μm to 2 μm with increasing CNTs content. The higher thickness detected for CNT rich samples is related to the increasing viscosity of the precursor solution. As already found for the CNT free samples the tensile pull-off force of detaching was for all coatings higher than 8 MPa. Furthermore, contact angle measurements have shown that no significant variation of the H_2O wetability (76±1° - 81±1°) of the film surface occurred with increasing CNTs content.

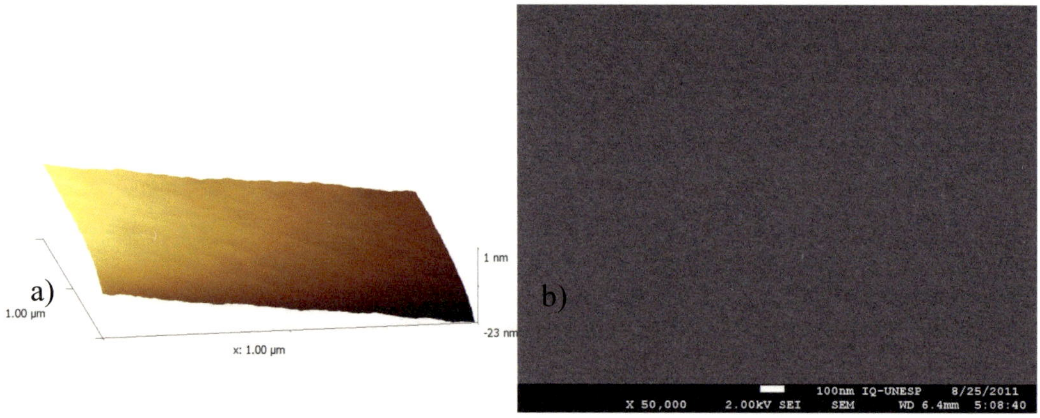

Fig. 10. Flat and featureless surface observed by a) AFM and b) FEG-SEM of the hybrid film prepared with a $[C_{CNT}]/[C_{CNT}+ Si]$ molar concentration of 5% (M4_5).

The verification of the degree of functionalization of the CNTs by carboxyl groups was performed using XPS. Figure 11 shows the C 1s core level spectra of as-received CNTs (Fig. 11a) and two functionalized samples after 4 h (Fig. 11b) and 8 h (Fig. 11c) of acidic treatment. The strong increase of the carboxyl component at about 288.8 eV evidences the increasing degree of functionalization with treatment time. From the intensity ratio of the O-C=O/C-C peak area a high functionalization degree of 0.15 and 0.35 was determined for the samples submitted to 4 h and 8 h of acid treatment, respectively. It cannot be excluded that, in the case of the latter sample, the high functionalization degree and the applied ultrasonic treatment of the precursor solution could lead to a significant fragmentation of the nanotubes (Voiry et al., 2011). Therefore, MWCNTs prepared after 4 h were used for the final formulation of the nanocomposit coating. Furthermore, it is important to point out that for the fitting of the aromatic carbon peak (C-C at 284.4 eV) an asymmetric Doniach-Sunjic function was used to account for the high density of states of multiwall carbon nanotubes at the Fermi level (Briggs & Seah, 1990). Thus it is possible to reduce the intensity of the carbon-oxygen bonds of the C 1s spectra in a way that the amount of these bonds is compatible with the oxygen concentration, determined from the integrated intensity of the O 1s peak.

Structural changes induced by the incorporation of increasing quantities of CNTs in the hybrid network were evaluated from the evolution of fitted XPS C 1s, O 1s and Si 2p core level spectra (Fig. 12). The samples were prepared with MMA/MPTS molar ratios of 4 (M4) containing carbon nanotubes with a $[C_{CNT}]/[C_{CNT}+ Si]$ molar concentration of 0.1% (M4_01) 1.0% (M4_1) and 5% (M4_5). The comparison of the C 1s spectral intensities (CPS) shows that the M4_1 and M4_5 sample contain a higher fraction of carboxyl groups than the M4 hybrid. The intensity of this sub-peak is comparable to that of the M10 sample, having a larger portion of these groups ester due to the higher content of the organic PMMA phase. This fact can be seen even clearer from the increase of the O-C=O component in the O 1s spectra. It is suggested that these additional carboxylic structures promote covalent bonds between the inorganic network and CNTs.

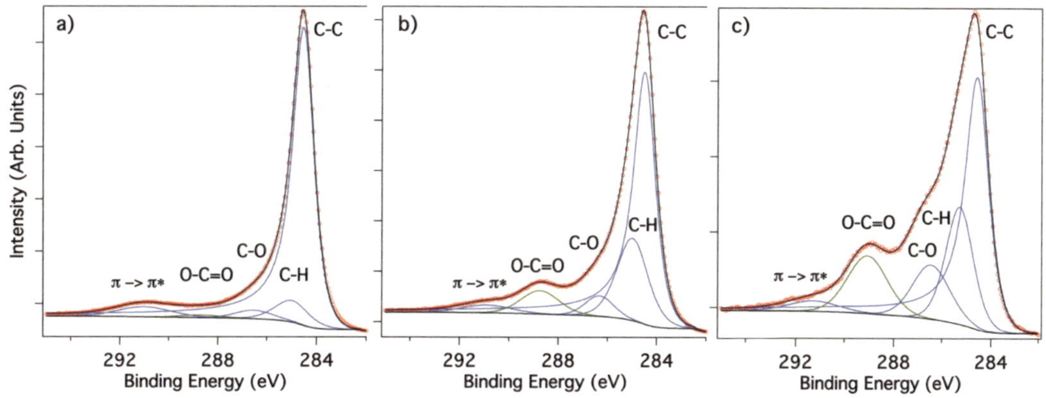

Fig. 11. Fitted XPS C1s core level spectra for a) as-received CNTs and b) after 2 h and c) 4 h of carboxyl functionalization by acidic treatment.

Figure 13 shows ^{29}Si NMR spectra for unsupported hybrid samples prepared at a MMA/MPTS ratio of 4 containing an increasing molar concentration of CNTs of 0% (M4), 0.1% (M4_01), 1.0% (M4_1) and 5% (M4_5). The degree of condensation extracted for the samples from the spectra were: 79.7±0.5%, 73.8±0.5%, 73.1±0.5% and 78.4±0.5%. Surprisingly, exactly for the sample containing the highest concentration of CNTs a similar degree of connectivity was found as for the M4 reference sample. Considering that the atomic concentration of silicon in the M4 hybrid, is close to 10 at.% (see XPS results), the carbon content related to CNTs of the sample M4_5 can be estimated to be about 0.5 at.%. This means that this considerable amount of CNTs is not affecting the high connectivity of the inorganic backbone structure of the M4 hybrid. In contrast, M_01 and M_1 hybrid show about 5% lower degrees of condensation, indicating the formation of a possibly less stable structure.

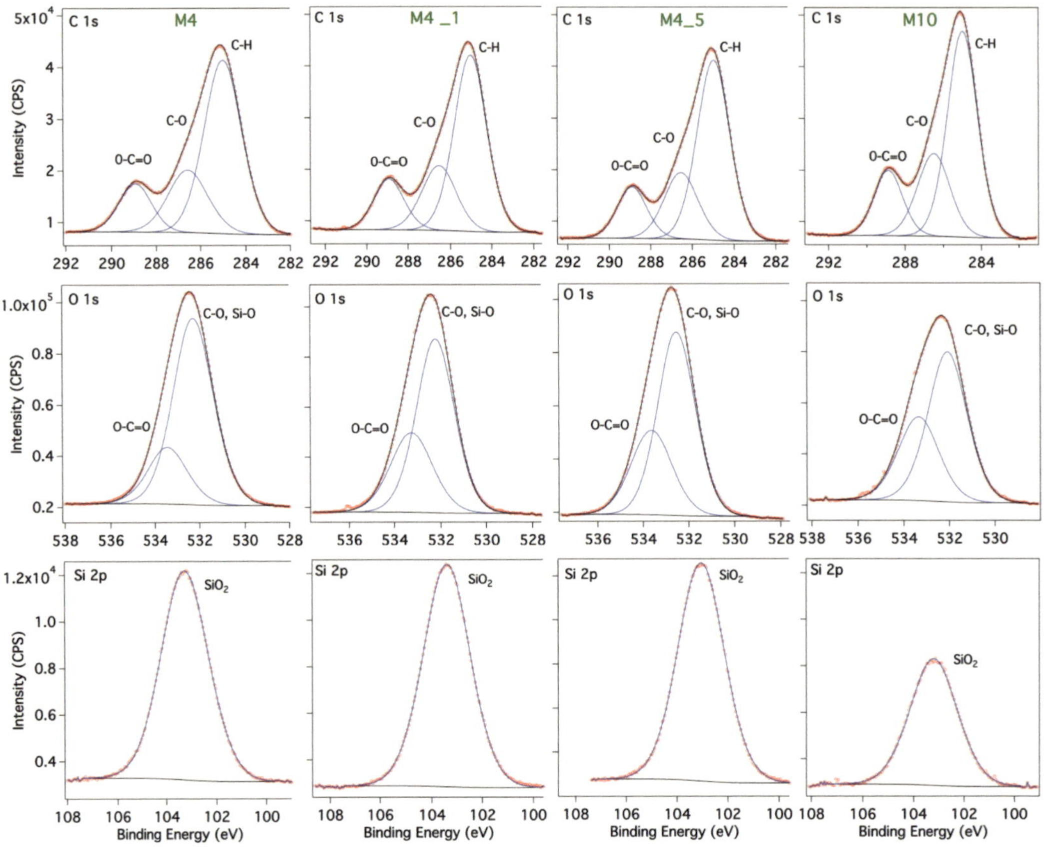

Fig. 12. Fitted XPS, C 1s, O 1s and Si 2p spectra of the hybrids prepared with MMA/MPTS ratios of 10 (M10) and 4 (M4) containing CNTs with [C_{CNT}]/[C_{CNT}+ Si]molar concentrations of 1.0% (M4_1) and 5% (M4_5).

The TGA and DTG curves of hybrids containing increasing amount of CNTs are shown in Figure 14. It is interesting to note, that in comparison to the M4 sample, the M4_5 hybrid displays two additional events at about 280 °C an 410 °C, the latter also observed for the M4_1 sample. The 410 °C event marks the thermal stability limit of these samples, a value found also for the M2 hybrid. As the intensity of the 280 °C event increases with the CNTs content, it is quite probable that this might be related to the liberation of functional groups, like carboxyls, attached to nanotube walls. From the TG/DTG results it can be concluded that the inclusion of higher quantities of CNTs maintains the thermal stability of the samples, indicating for these hybrids that the CNTs reinforce the structure possibly by the presence of covalent bonds with siloxane groups of the hybrid network.

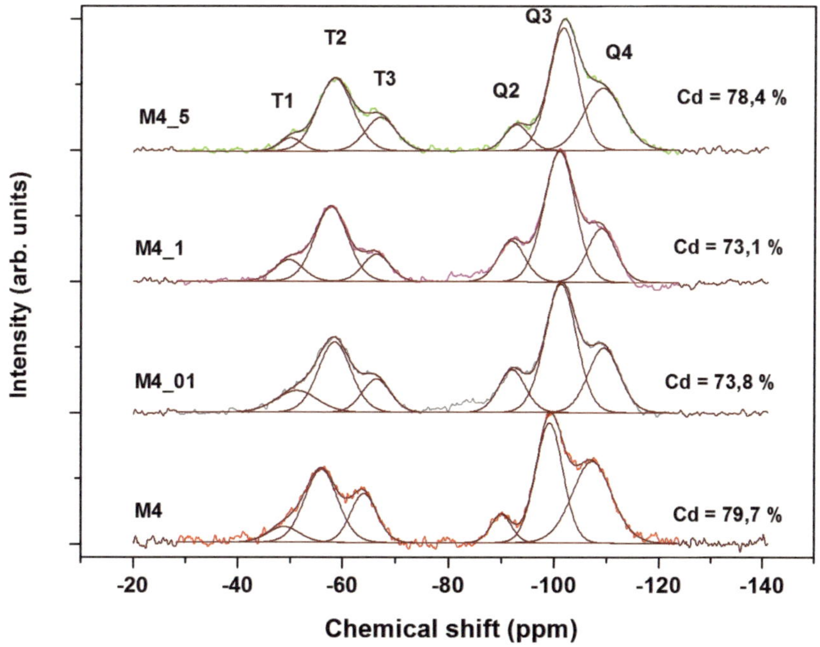

Fig. 13. ^{29}Si NMR spectra of hybrids prepared with a MMA/MPTS a ratio of 4 containing CNTs with $[C_{CNT}]/[C_{CNT}+ Si]$ molar concentrations of 0% (M4), 0.1% (M4_01), 1.0% (M4_1) and 5% (M4_5).

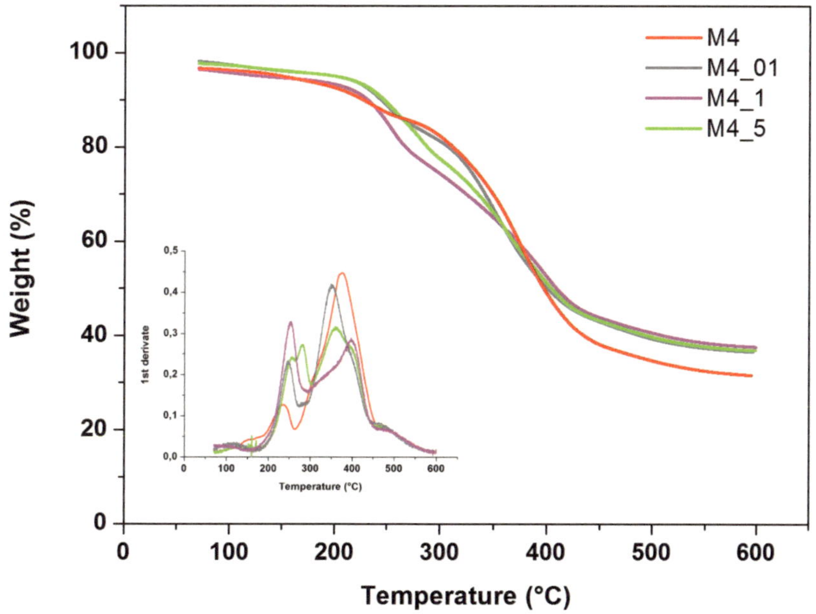

Fig. 14. TGA and DTG (insert) curves of hybrids prepared with a MMA/MPTS a ratio of 4 containing CNTs with $[C_{CNT}]/[C_{CNT}+ Si]$ molar concentrations of 0% (M4), 0.1% (M4_01), 1.0% (M4_1) and 5% (M4_5).

3.2.2 Corrosion analysis

The electrochemical characteristics of hybrids coatings containing different concentrations of carbon nanotubes was investigated using electrochemical impedance spectroscopy curves, recorded after 1 day of immersion in acidic and saline solution. Figure 15 shows the complex plane impedance and the Bode plots (log |Z| and θ vs. log f) for the M4 reference sample and the CNTs containing hybrid films M4_01, M4_1 and M4_5. Compared to the M4 film, the M_01 an M_5 films show a very similar characteristic in the saline solution. Only the M4_1 coating displays a second LF time constant, related to oxidized steel surface due to the penetration of the electrolyte to the film/steel interface. The beginning degradation can be associated with a less stable structure of this coating as suggested by the lowest degree of condensation, observed by NMR. The obtained model parameters are displayed in Table 3 show a slightly lower corrosion resistances of the M4_1 coating. A quite different behavior was observed after immersion of the samples into the acidic medium. While for the CNT coatings the corrosion resistance remains essentially unchanged a drop of the impedance, of more than one order of magnitude, was detected for the M4 coating. The inferior performance in acidic solution is also indicated by the appearance of the LF time constant, marking the initiation of film failure. This is also evidenced by the lower corrosion resistance of M4 compared to that of M4_5 sample (see Table 3). The obtained results suggest a superior performance of the CNT coatings at low pH and in the case of M4_1 coating a higher susceptibility for elevate concentration of chorine ions, present in the saline solution.

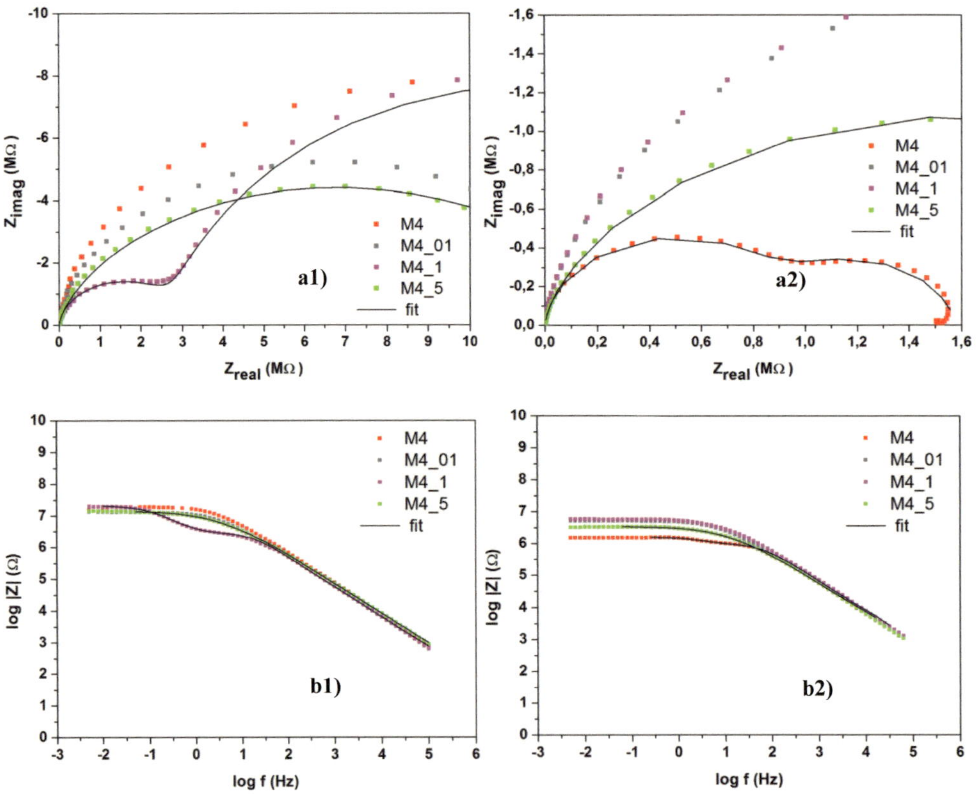

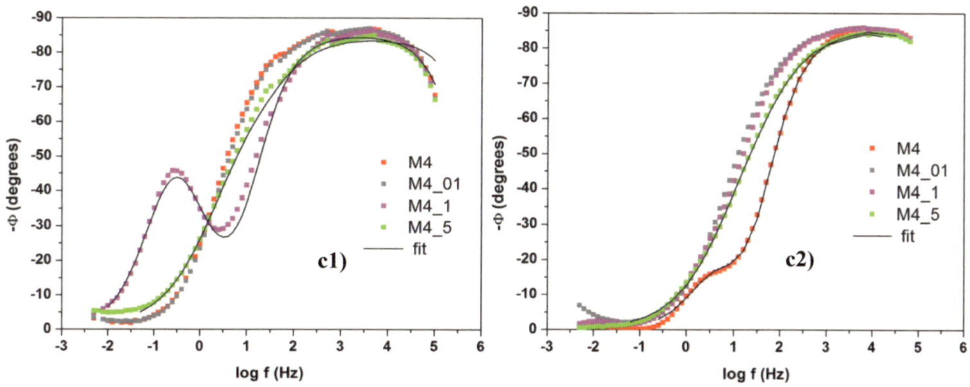

Fig. 15. Complex plane impedance (a), modulus impedance (b) and phase (c) plots of bare carbon steel and hybrids prepared with a MMA/MPTS a ratio of 4 containing CNTs with $[C_{CNT}]/[C_{CNT}+ Si]$ molar concentrations of 0% (M4), 0.1% (M4_01), 1.0% (M4_1) and 5% (M4_5) after exposure of 1 day in (1) unstirred and naturally aerated 3.5 % NaCl and (2) 0.05 mol L^{-1} NaCl + 0.05 mol L^{-1} H_2SO_4 solutions. (The small discontinuities of θ vs. log f mark the change of the measurement range of the instrument.)

Sample	M4_5 saline	M4_1 saline	M4_5 acid	M4 acid
χ^2 (10^{-3})	7.2	6.6	0.4	0.9
R_{sol} (Ω cm^2)	27.6 (10.4)*	29.3 (12.2)*	19.5 (8.9)	20.0 (11.2)*
CPE_1 (10^{-3} μF cm^{-2} $s^{\alpha-1}$)	3.95 (3.9)	4.34 (2.4)	39.8 (6.9)	129 (9.9)
n_1	0.93 (0.4)	0.94 (0.3)	0.59 (3.4)	0.89 (4.3)
R_1 (10^6 Ω cm^2)	3.2 (14.9)	2.7 (1.8)	2.51 (0.2)	0.60 (4.7)
CPE_2 (10^{-3} μF cm^{-2} $s^{\alpha-1}$)	18.2 (1.9)	85.2 (1.9)	5.18 (4.9)	3.94 (3.4)
n_2	0.56 (1.5)	0.86 (1.5)	0.93 (0.6)	0.94 (0.4)
R_2 (10^6 Ω cm^2)	1.97 (2.5)	1.63 (2.5)	1.07 (12)	0.94 (1.7)

*Error (%)

Table 3. Parameters of the electrical equivalent circuit for sample M4_5, M4_1 and M4 after 1 day immersion in neutral 3.5% NaCl aqueous and 0.05 mol L^{-1} NaCl + 0.05 mol L^{-1} H_2SO_4 solution.

The higher anti-corrosion efficiency of CNTs containing coatings in acidic medium was confirmed by the comparison of the potentiodynamic polarization curves of the M4 film with those of M4_01, M4_1 and M4_5 coatings, displayed in Figure 16. After 2 h of immersion in acidic NaCl solution the current density of the M4 film is almost one order of magnitude lower than the 10^{-9} Acm^{-2} observed for the CNTs containing coatings (Fig. 16b). Also in the case of exposure to the neutral saline solution (Fig. 16a) the polarization curves of the CNTs samples are very similar showing slightly higher current densities than sample M4, with value of about $8x10^{-10}$ Acm^{-2}. It should be noted, that for all tested films no rupture of the coating was observed in the positive potential branch of the overpotentials of up to 1 V. According to results obtained by scanning electron microscopy and atomic force microscopy the surface of all CNTs containing samples is very smooth with a surface RMS rougness of less than 0.2 nm. Consequently an increase of the anodic current due to larger cathode surface area on increasing doping was not observed. In contrary, Figure 16 shows that the current density of M4_5 over the anodic potential branch is the lowest in both environments. According to the electrochemical results, the performance against corrosion in neutral saline medium can be established in the following order: M4 > M_5 ≈ M_01 > M_1 > steel. In the case of acidic NaCl solution the established sequence is: M_5 ≈ M_01 ≈ M_1 > M4 > steel. The latter result indicates that the presence of carbon nanotubes in the hybrid structure increases the inert character of the hybrid coatings at low pH values.

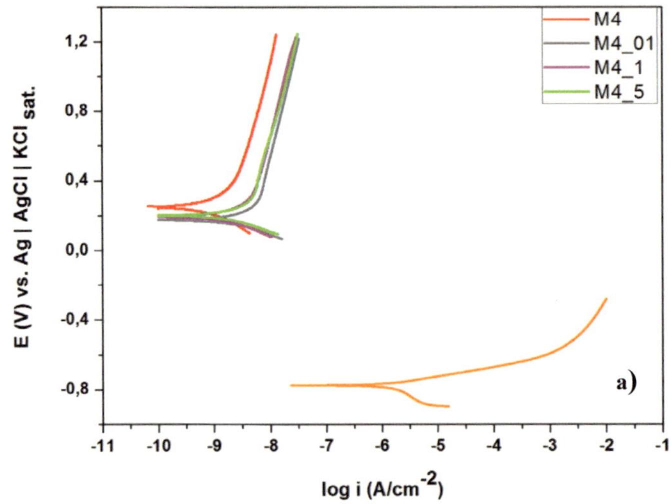

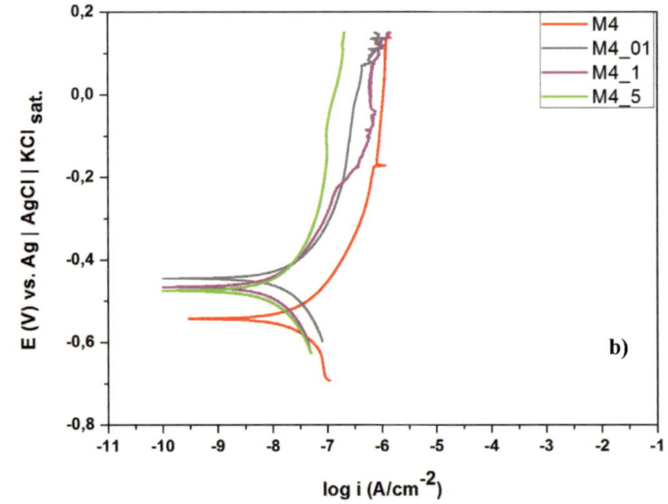

Fig. 16. Potentiodynamic polarization curves of bare carbon steel and hybrids prepared with a MMA/MPTS a ratio of 4 containing CNTs with $[C_{CNT}]/[C_{CNT}+ Si]$ molar concentrations of 0% (M4), 0.1% (M4_01), 1.0% (M4_1) and 5% (M4_5) recorded after 2 h immersion in a) unstirred and naturally aerated 3.5 % NaCl and b) 0.05 mol L^{-1} NaCl + 0.05 mol L^{-1} H_2SO_4 solution.

4. Conclusions

Smooth, crack-free, adherent and optically transparent siloxane-polymethyl methacrylate hybrid coatings were deposited onto carbon steel from sol prepared by acid-catalyzed hydrolytic polycondensation of TEOS and MPTS mixtures at TEOS/MPTS molar ratio of 2, followed by radical polymerization of MMA under varying MMA/MPTS ratio of 2, 4, 8 and 10. XPS results confirmed the increasing proportion of the polymeric phase and NMR results showed that the degree of polycondensation of silicon species had a maximum of about 84% for a MMA/MPTS ratio of 8, while the extent of organic polymerization was essentially unaffected by the variation of MMA. The excellent corrosion protection efficiency of the coatings, with a corrosion resistance of up to 1 GΩ and current density of less than 10^{-10} Acm^{-2} determined by electrochemical measurements for sample M8 is closely related to the ramified structure of siloxane cross-link nodes of the hybrid network. For this sample, no corrosion-induced changes were observed by XPS for up to 18 days of immersion in a NaCl solution. The very high breakdown potential of this coating and the low current density even for high anodic potentials are also indicative for the exceptional performance of these coatings against corrosion. Thus the high polycondensation degree of the inorganic backbone results in the formation of a dense hybrid coating that acts as an efficient diffusion barrier, protecting carbon steel against corrosion, principally in neutral chloride medium.

XPS and TG/DTG results obtained for carbon nanotubes containing hybrids, indicate that the incorporated CNTs increase the stability of the hybrid network possibly by formation of covalent bonds with siloxane groups. Electrochemical measurements have shown similar anti-corrosion characteristics than the undoped samples in neutral saline environment, while a superior performance was detected in acidic NaCl media, due to more inert character of the hybrid structure reinforced by carbon nanotubes.

5. Acknowledgements

The authors like to thank for the financial support for this work provided by CAPES, CNPq and FAPESP.

6. References

Betova, I.; Bojinov, M.; Laitinen, T.; Makela, K.; Pohjanne,P.; Saario, T. (2002). The Transpassive Dissolution Mechanism of Highly Alloyed Stainless Steels: I. Experimental Results and Modeling Procedure. *Corrosion Science*, Vol.44, No.12, (March 2002), pp. 2675 - 2697, ISSN 0010-938X

Bhattacharyya, S.; Das, M.B.; Sarkar. S. (2008). Failure Analysis of Stainless Steel Tubes in a Recuperator Due to Elevated Temperature Sulphur Corrosion. *Engineering Failure Analysis*, Vol.15, No.6, (September 2008), pp. 711 - 722, ISSN 1350-6307

Briggs, D. & M.P. Seah M.P. (1994), *Practical Surface Analysis: Auger and X-ray Photoelectron Spectroscopy*, John Wiley & Sons, ISBN 0-471-92081-9, Chichester, England

Brinker, C. J. & Scherrer, G. W. (1990). *Sol-gel science*, Academic Press, ISBN 13:978-0-12-134970-7, New York

De Graeve, I.; Vereecken, J.; Franquet, A.; Van Schaftinghen, T.; Terryn, H. (2007). Silane Coating of Metal Substrates: Complementary Use of Electrochemical, Optical and

Thermal Analysis for the Evaluation of Film Properties. *Progress in Organic Coating,* Vol.59, No.3, (June 2007), pp. 224 - 229, ISSN 0300-9440

Droppa Jr. R.; Hammer, P.; Carvalho A.C.M.; Dos Santos, M. C.; Alvarez, F. (2001). Incorporation of Nitrogen in Carbon Nanotubes. *Journal of Non-Crystalline Solids,* Vol.299-302, part 2, (April 2002), pp. 874-879, ISSN 0022-3093

Eder. D. (2010). Carbon Nanotube Inorganic Hybrids. *Chemical Reviews,* Vol.110, No.3, (March 2010), pp 1348–1385, ISSN 0009-2665.

Fedrizzi L.; Rodriguez F.J.; Rossi S.; Deflorian F.; Di Maggio R. (2001). The Use of Electrochemical Techniques to Study the Corrosion Behavior of Organic Coatings on Steel pretreated with Sol-Gel Zirconia Films. *Electrochimica Acta,* Vol.46, No.24-25, (August 2001), pp. 3715 - 3724, ISSN 0013-4686

Feliu V.; González J.A.; Andrade C.; Feliu S. (1998). Equivalent circuit for modeling the steel-concrete interface. II. Complications in applying the stern-geary equation to corrosion rate determination. Corros. Sci. Vol.40, No.6, pp. 995-1006, ISSN 0010-938X.

Hammer, P.; Schiavetto, M. G.; dos Santos, F. C.; Pulcinelli, S. H.; Benedetti, A. V.; Santilli, C. V. (2010). Improvement of the Corrosion Resistance of Polysiloxane Hybrid Coatings by Cerium Doping. *Journal of Non-Crystalline Solids,* Vol.356, No. 44-49, (October 2010), pp. 2606 - 2612, ISSN 0022-3093.

Han,Y.H.; Taylor, A.; Mantle, M.D.; Knowles, K.M. (2007). UV curing of organic–inorganic hybrid coating materials. *Journal of Sol–Gel Science and Technology,* Vol.43, No.1, (January 2007), pp. 111 - 123, ISSN 0928-0707

Harreld, J.H.; Esaki, A.; Stucky, G.D. (2003). Low-Shrinkage, High-Hardness, and Transparent Hybrid Coatings: Poly(methyl methacrylate) Cross-Linked with Silsesquioxane. *Chemistry of Materials,* Vol.15, No.18, (July 2003), pp. 3481-3489, ISSN 0897-4756

Innocenzi, P.; Brusatin, G.; Licoccia, S.; Di Vona, M.L.; Babonneau, F.; Alonso, B. (2003). Controlling the Thermal Polymerization Process of Hybrid Organic-Inorganic Films Synthesized from 3-Methacryloxy-propyltrimethoxysilane and 3-Aminopropyltriethoxysilane. *Chemistry of Materials,* Vol.15, No.25, (November 2003), pp.4790 - 4797, ISSN 0897-4756

José, N. M.; Prado, L. A. S. A. (2005). Materiais híbridos orgânico-inorgânicos: preparação e algumas aplicações. *Química Nova,* v.28, No.2, (November 2004), pp. 281-288, ISSN 0100-4042

Khare, R. & Bose, R. (2005). Carbon Nanotubes Based Composits - A Review. *Journal of Minerals & Materials Characterization & Engineering,* Vol.4, No.1, pp. 31-46, ISSN 1539-2511

Kim, M.; Hiong, J.; Hong, C.K.; Shim, S.E. (2009). Preparation of Silica-Layered Multi-Walled Carbon Nanotubes Activated by Grafting of Poly(4-vinylpyridine). *Synthetic Metals,* Vol.159, No.1-2, (January 2009), pp. 62-68, ISSN 0379-6779

Landry, C. J. T.; Coltrain, B.K.; Brady, B.K. (1992). In Situ Polymerization of Tetraethoxysilane in poli(methyl methacrylate): Morphology and Dynamic Mechanical Properties. *Polymer,* Vol.33, No.7, (February 1991), pp. 1486 - 1494, ISSN 0032-3861

Lopez, D.A; Rosero-Navarro, N.C.; Ballarre, J.; Durán, A.; Aparicio, M.; Ceré, S. (2008). Multilayer Silica-methacrylate Hybrid Coatings Prepared by Sol–Gel on Stainless

Steel 316L: Electrochemical Evaluation. *Surface & Coatings Technology*, Vol.202, No.10, (February 2008), pp. 2194 - 2201, ISSN 0257-8972

Masalski, J.; Gluszek, J.; Zabrzeski, J.; Nitsch, K.; Gluszek, P. (1999). Improvement in Corrosion Resistance of the 3161 Stainless Steel by Means of Al_2O_3 Coatings Deposited by Sol-Gel Method. *Thin Solid Films*, Vol.349, No.1, (July 1999), pp. 349 186 - 190, ISSN 0040-6090

Messaddeq, S. H.; Pulcinelli, S. H.; Santilli, C. V.; Guastaldi, A. C.; Messaddeq, Y. (1999). Microstructure and Corrosion Resistance of Inorganic-Organic (ZrO_2-PMMA) Hybrid Coating on Stainless Steel. *Journal of Non-Crystalline Solids*, Vol.247, No.2, (December 1999), pp. 164-170, ISSN 0022-3093

Metroke, T. L.; Kachurina, O.; Knobbe, E. T. (2002). Spectroscopic and Corrosion Resistance Characterization of GLYMO-TEOS Ormosil Coatings for Aluminum Alloy Corrosion Inhibition. *Progress in Organic Coatings*, Vol.44, No.4, (August 2002), pp. 295-305, ISSN 0300-9440

Nazeri, A; TrzaskomaPaulette, P.P.; Bauer, D. (1997). Synthesis and Properties of Cerium and Titanium Oxide Thin Coatings for Corrosion Protection of 304 Stainless Steel. *Journal of Sol–Gel Science and Technolology*, Vol.10, No.3, (November 1997), pp. 317 - 331, ISSN 0928-0707

Pepe, A.; Aparicio, M.; Cere, S.; Duran, A. (2004). Preparation and Characterization of Cerium Doped Silica Sol-Gel Coatings on Glass and Aluminum Substrates. *Journal of Non-Crystalline Solids*, Vol.348, No.15, (November 2004), pp. 162-171, ISSN 0022-3093

Ryan, M.R.; Williams, D.E. ; Chater, R.J.; Hutton, B.M ; McPhail, D.S. Why Stainless Steel Corrodes. (2002). *Nature*, Vol.415, No.6873, (February 2002), pp. 770- 774, ISSN 0028-0836

Sanchez, C. & Lebeau, B. (2004). Optical Properties of Functional Hybrid Organic-Inorganic Nanocomposites, In: *Functional Hybrid Material*, Gómez-Romero, P., pp. (122-168), Oxford: Elsevier, ISBN 3-527-30484-3, Germany

Saravanamuttu, K.; Du, X. M.; Najafi, S.I.; Andrews, M.P. (1998). Photo-induced Structural Relaxation and Densification in Sol-Gel Derived Nanocomposite Thin Films; Implications in Integrated Optics Device Fabrication. *Canadian Journal of Chemistry / Revue Canadienne de Chimie*, Vol.76, No.11, (November 1998), pp.1717 - 1729, ISSN 1480-3291

Sarmento, V. H. V.; Frigerio, M.R.; Dahmouche, K.; Pulcinelli, S. H.; Santilli, C. V. (2006). Evolution of Rheological Properties and Local Structure During Gelation of Siloxane-Polymethylmethacrylate Hybrid Materials. *Journal of Sol–Gel Science and Technology*, Vol.37, No.3, (February 2006), pp. 179 - 174, ISSN 0928-0707

Sarmento, V.H.V.; Schiavetto, M.G.; Hammer, P.; Benedetti, A.V.; Fugivara, C.S.; Suegama, P.H.; Pulcinelli, S.H.; Santilli, C.V. (2010). Corrosion Protection of Stainless Steel by Polysiloxane Hybrid Coatings Prepared Using the Sol-Gel Process. *Surface & Coatings Technology*, Vol.204, No.16-17, (February 2010), pp. 2689-2701, ISSN 0257-8972

Stern, M.; Geary A.L. (1957). Electrochemical polarization. Journal of Electrochem. Soc., Vol.104, No.1, pp. 56-63, ISSN 1945-7111.

Suegama, P.H.; de Melo, H.G.; Benedtti, A. V.; Aoki, I.V. (2009). Influence of Cerium (IV) Ions on the Mechanism of Organosilane Polymerization and on the Improvement of its Barrier Properties. *Electrochimica Acta*, Vol.54, No.9, (March 2009), pp. 2655-2662, ISSN 0013-4686

Suegama, P.H.; Espallargas, N.; Guilemany, J. M.; Fernández, J.; Benedetti, A. V. (2006). Electrochemical and Structural Characterization of Treated Cr_3C_2-NiCr Coatings. *Journal of Electrochemical Society,* Vol.153, No.10, pp. B434-B445, ISSN 1945-7111

Tadanaga K. ; Ellis B. ; Seddon A.B. (2000). Near- and Mid-Infrared Spectroscopy of Sol-Gel Derived Ormosil Films for Photonics from Tetramethoxysilane and Trimethoxysilylpropylmethacrylate. *Journal of Sol–Gel Science and Technology,* Vol.19, No.1 - 3, (December 2000), pp. 687 - 690, ISSN 0928-0707

Tsutsumi, Y.; Nishikata, A.; Tsuru, T. (2007). Pitting corrosion mechanism of Type 304 stainless steel under a droplet of chloride solutions. *Corrosion Science,* Vol. 49, No.3, (October 2006), pp. 1394 - 1407, ISSN 0010-938X

Vasconcelos, D.C.L.; Carvalho, J.A.N.; Mantel, M.; Vasconcelos, W.L. (2000). Corrosion Resistance of Stainless Steel coated with Sol-Gel Silica. *Journal of Non–Crystalline Solids,* Vol.273, No.1-3, (August 2000), pp. 135 – 139, ISSN 0022-3093

Voiry, D.; Vallés, C.; Roubeau, O.; Pénicaud, A. (2011). Dissolution and alkylation of Industrially Produced Multi-Walled Carbon Nanotubes. *Carbon,* Vol.49, No.1, (January, 2011), pp. 170-175, ISSN 0008-6223.

Wang, Y. & Bierwagen, G.P.P. (2009). A New Acceleration Factor for the Testing of Corrosion Protective Coatings: Flow-Induced Coating Degradation. *Journal of Coatings Technology and Research,* Vol.6, No.4, (December 2009), pp. 429-436, ISSN 1547-0091

Zandi-Zand, R.; Ershad-Langroudi, A.; Rahimi, A. (2005). Organic-Inorganic Hybrid Coatings for Corrosion Protection of 1050 Aluminum Alloy. *Journal of Non-Crystalline Solids,* Vol.351, No.14-15, (May 2005), pp. 1307-1311, ISSN 0022-3093

Zheng, S. & Li, J. (2010). Inorganic–Organic Sol Gel Hybrid Coatings for Corrosion Protection of Metals. *Journal of Sol-Gel Science and Technology,* Vol.54, No.2, (May 2010), p. 174-187, ISSN 0928-0707

Zheludkevich, M.L.; Salvado, I.M.; Ferreira, M.G.S. (2005). Sol-gel Coatings for Corrosion Protection of Metals. Journal of Materials Chemistry, Vol.15, No.18, (July 2005), pp. 5099-5111, ISSN 0959-9428